AF863102

Power Systems

For further volumes:
http://www.springer.com/series/4622

Benoît Robyns · Bruno Francois
Philippe Degobert · Jean Paul Hautier

Vector Control of Induction Machines

Desensitisation and Optimisation Through Fuzzy Logic

Springer

Benoît Robyns
Hautes Etudes d'Igénieur
13 rue de Toul
59046 Lille cedex
France

Philippe Degobert
Arts et Métiers ParisTech
8 Boulevard Louis XIV
59046 Lille cedex
France

Bruno Francois
Ecole Centrale de Lille
Boulevard Paul Langevin
59651 Villeneuve d'Ascq cedex
France

Jean Paul Hautier
Arts et Métiers ParisTech
151 Boulevard de l'Hôpital
75013Paris cedex
France

ISSN 1612-1287
ISBN 978-0-85729-900-0
DOI 10.1007/978-0-85729-901-7
Springer London Heidelberg New York Dordrecht

e-ISSN 1860-4676
e-ISBN 978-0-85729-901-7

British Library Cataloguing in Publication Data
A catalogue record for this book is available from the British Library

Library of Congress Control Number: 2011944633

© Springer-Verlag London 2012
MATLAB is a registered trademark of The Mathwork, Inc., 3 Apple Hill Drive, Natick, Massachusetts, United States, 01760.
Apart from any fair dealing for the purposes of research or private study, or criticism or review, as permitted under the Copyright, Designs and Patents Act 1988, this publication may only be reproduced, stored or transmitted, in any form or by any means, with the prior permission in writing of the publishers, or in the case of reprographic reproduction in accordance with the terms of licenses issued by the Copyright Licensing Agency. Enquiries concerning reproduction outside those terms should be sent to the publishers.
The use of registered names, trademarks, etc., in this publication does not imply, even in the absence of a specific statement, that such names are exempt from the relevant laws and regulations and therefore free for general use.
The publisher makes no representation, express or implied, with regard to the accuracy of the information contained in this book and cannot accept any legal responsibility or liability for any errors or omissions that may be made.

Printed on acid-free paper

Springer is part of Springer Science+Business Media (www.springer.com)

Foreword

It is a real pleasure to welcome the electrical engineering readership to the English version of this timely and comprehensive monograph from Professors Robyns, François, Degobert and Hautier, as the first self-contained approach to the desensitization of induction-machine vector-control systems against parameter uncertainties.

There are several structural features that conspire in making this book a valuable addition to the literature. First, the use of Causal Ordering Graphs formalism allows a systematic design and a standardization of vector control techniques based on indirect rotor-flux orientation.

Second, the theoretical analysis of parameter sensitivity is applied to various vector control strategies, and leads to their robustness-related combination in order to desensitize the induction-machine vector controls, as well as to optimize the choice of observer gains for many operating conditions.

Third, an original fuzzy logic supervisor is developed from a theoretical study of sensitivity to minimize that part of the parameter uncertainty due to induction machine models, and to manage the remaining uncertainty part that cannot be removed.

Fourth, practical implementation aspects of desensitized induction-machine vector-control systems are thoroughly addressed by two industrial applications: one relying on fuzzy logic supervisor-based combination of vector control strategies for an induction machine-driven flywheel energy storage device associated to a hybrid wind/diesel power generation assembly in order to smooth the wind-generated irregular power; and another hinging on the torque control of a traction induction machine performed with a reduced-order flux observer optimized through fuzzy logic in order to mitigate errors on the torque estimate due to electrical parameter uncertainties, especially during the start-up period.

Skillfully written by distinguished researchers in the field, this highly topical and intriguing book is primarily aimed at advanced electrical engineering students and practitioners engaged in the increasingly complex area of induction-machine high-performance control systems.

Romania

Prof. Mircea M. Radulescu
Faculty of Electrical Engineering
Technical University of Cluj-Napoca

Preface

Electricity as a vector of future energy

For the general public, the second half of the twentieth century is considered, in terms of technology, as the age of electronics and electronic applications. From transistors to integrated circuits, microelectronics and nowadays nanoelectronics have given rise to inventions and devices that have completely changed our means of communication and information: television, cellular phone, personal computers, CDs, DVDs, communication satellites and the Internet have become so overwhelming that our society is often referred to as the society of information and communication. The survey carried out by the trustworthy association of electricians, electronics engineers and radio electricians (IEEE), in the November 2004 issue of the journal *Spectrum*, is therefore not surprising at all. Out of the 40 specialists surveyed in terms of research priority and major technological evolutions over the next decade, 38 mainly mentioned Internet, telecommunications, nanoelectronics, nano and biotechnologies among others. Only one of them considered energy products as a priority for our future. This is striking if we consider the current revolutions taking place in the energy field: scarcity and rocketing cost of raw materials, market liberalization, high expectancies for engineers to accomplish miracles—transformation of the laws of physics to constraint currents to follow the laws of the market, and Kirchhoff's laws not considered as obsolete anymore—and some European regions on the industrial decline due to a lack of foresight in the energy field.

Whatever the perspectives may be in terms of communication society, the twenty first century will be characterized by deep changes in our energy production and consumption. Be it due to the scarcity of some raw materials or environmental issues, we are about to enter a new era in which all potentially useful sources of energy will be needed and the consumption of available energy will have to be as rational as possible.

In this context, electricity will become a major vector in terms of energy production be it from natural resources such as wind, tides, swell, river flow, sun or from sources of nuclear origin with nuclear fusion as a long-run objective.

Electricity will also become a major vector in terms of rational use of produced energy. Cogeneration, wind mills, hydropower plants, hybrid vehicles and high-speed rail transport as a competitor of air transport, are among the many elements that already embody this evolution. Electrical engineering has evolved increasingly in the last 30 years, thus making these applications possible. Although they were not very apparent, research works in the electrical engineering field were numerous and innovative in those years. Material evolutions, progress in terms of computer-aided design, the development of power electronics and control and setting techniques now enable the design of new electromechanical conversion systems, innovative both in terms of performance and fields of application.

The induction machine is a striking example of this evolution. Dozens of years ago, this network-powered machine still seemed a cost-effective and robust machine, yet only capable of driving at a quasi constant speed load that did not require frequent starts and stops to gain an acceptable energy yield. Nowadays, powered with a voltage inverter with pulse-wave modulated control, it has become the reference machine for variable speed drive going from kW to MW. By coupling up its inverter with a vector control, this machine even offers dynamics and high torque accuracy, which is well suited to applications requiring high-speed traction. Equipped with a wound rotor, network connected, thanks to its stator and powered with a rotor converter, it has also become a generator well suited to the production of electrical energy by windmills since it enables the turbine to adapt its speed based on that of the wind.

If one tries, however, to reach its theoretical limits in terms of accuracy and setting, the induction machine remains hard to control. Indeed, one cannot generally measure all the quantities that are specific to its state and the estimation (observation) of the non-measurable quantities is toughened by the evolution of its parameters due to temperature and saturation. All issues are therefore far from being solved.

Vector Control is a Standard

The "vector control" of alternating current machines, heritage of Blaschke works in the early seventies, has become a settled standard. This method has largely been used and deepened by researchers. When technologies became mature enough (micro-processors for real-time calculations, power electronics for cost-effective inverters), it gave rise to wide industrial ranges in the early nineties. The possibility to give the induction machine dynamic performance similar to direct-current motors (the reference in that field) then appeared. It enabled benefit from two major advantages: an economical one—the induction machine being the least expensive actuator—and a technical one—from its absence of a mechanical commutator. The continuous decrease in the cost of permanent magnets may question this hierarchy in the future, putting induction machines at the front of the stage again. This machine is the ideal one, yet it is considered as too expensive for wide-scale applications.

The vector control is a "standard" as it is widely used: for instance, every TGV (High Speed Train) passenger benefits from it. However, the adjective "standard" does not imply that this method prioritizes a unique solution. The "vector control" sets the basis for "basic" mathematical methods and models. It leads to various control architectures as it offers a decoupling between the "d axis" (the "flux axis", rotor flux in its classical version) and the "q axis" (the "torque axis"). People used to induction machine control using the Park transform know that very well. More generally, the vector control of the rotor flux of the induction machine is an intelligent transposition of the good properties of the synchronous machine, which itself was a transposition of the excellent properties of the direct current motor.

Although many books can be found both in French and English on induction machines control, the book written by Benoît Robyns, Bruno François, Philippe Degobert and Jean Paul Hautier differs from the others thanks to its originality, as shown in the following paragraphs.

Control, estimation, observation and robustness are many issues that need to be taken into account simultaneously

The following paradox must be emphasized: the vector control owes its good properties to the emphasis put on a particular variable, the rotor flux, that gives way to a "simple model" but usually this flux is a non-measurable internal variable. Simplicity is a relative concept that will be linked to a more detailed concept, that of "direct model", that needs to be "reversed". The guiding principle of this book is the following: how can the initial issue of control be linked with the corollary and even more important issue of "estimation and observation" of the rotor flux? The hinted difficulty behind all that has to do with "robustness". How can the control be "robust" despite many difficulties in determining the rotor flux? These concepts will be further explained in this book.

The first context has to do with control or "direct model to reverse model"

Some of this book's authors have been pioneers in the designing of controls directly deduced from modeling. They have highlighted the fact that physical models can lead to control laws if a structured method is applied. "Direct modeling" can be deduced from physical laws. From direct modeling, naturally, a certain type of "inverse modeling" will emerge, that is to say, control laws and implantable algorithms. This method, based on physical arguments and especially on energy, is excellent. It benefited from a rich past. Control engineers use it with their own and more abstract tools in, for example, the "input–output linearization with state reversion" or the "passive control", that give similar control architectures yet require a quite consistent mathematical background. The authors also made a fundamental and strategic choice. This choice consisted in adopting a graphic representation, as opposed to the approach adopted by many system theoreticians that used very abstract representations only, that is to say only equations far from the physical reality. The authors have developed their own graphic representation, called the C.O.G. "Causal Ordering Graph", which is very useful from a methodological point of view. This graphic representation naturally forces a control designer to structure his/her system correctly. It is legitimate to think that other graphic representations are equally acceptable. The "good old transfer functions" allow for a similar approach, if and only if the designer constrains himself/herself to structure them correctly to highlight the physical properties that give way to "direct models". This condition is not always met, far from it. Many functional diagrams published in the scientific literature are indistinct and therefore hardly usable in control design. This requires a "certain expertise", which some people in the scientific community already have. Electrical engineers who worked on controls were used to "adjusting" their equations to "decouple the d and q axes", for instance. The method given by the authors enables to justify, structure and systematizes these researches on decoupling. The "input–output standardization" of the automation enables to apply this property as well, but under a more abstract mathematical form that requires either powerful mathematical tools (for instance, the "Lie algebra") or a "very good intuitive sense". The pedagogical virtues of the C.O.G. are real and enable students to become familiar with fundamental concepts such as "causality", with its "rigid relations" (pictures by algebraic equations) and its "causal relations" (described with integro-differential equations). "Rigid relations" can be "inverted in open loop" but causal relations can only occur in "closed loop". Justifications for this can be given either using "physics vocabulary" (and energy concept) or "mathematical vocabulary" (properties of differential equations) or "automatic vocabulary" (properties of closed-loop systems). These different approaches (graphical, physical and mathematical) go together very well and offer a variety of approaches that can be used depending on the expectancies of the reader.

Direct and reverse models of the induction machine, considering vector control

The authors of this book have applied their method to the classical modeling of the induction machine in dynamic state in the various references used in "control": the "natural" three-phased reference—the one used in windings and from which Faraday's law in physics was derived—and the two-phased references either fixed or rotating (Park transform). The term "control" will need to be further clarified, as it covers activities other than the control itself. What is obtained is not "a" model but "a class" of models since two fundamental and strategic choices remain to be made: choosing the frame of reference and the state variables. The author's methodology systematically tackles these two issues and associates graphic representations (thanks to C.O.G.) with differential equation representations and state model representations, as is commonly done in automatic control engineering.

This book does not pretend, however, to cover all use cases of these models. An encyclopedia, tedious reading, would be needed to do so. This book has to do only with the vector control depending on rotor flux. The authors therefore clearly place themselves in a "control context". The model obtained has to be a "direct model easily reversible". This brings the authors to the concept of "maximum structure control", highly competitive yet extremely requiring. This "maximum structure control" leads to control algorithms that associate open-loop "compensations" (either additive or multiplicative) and closed-loop regulations. All state variables need to be known at all times. Hence the fundamental need for real-time computation of non-measurable variables. Naturally, this process leads to the concepts of controllers, estimators and observers, which will be explained further.

When using controller controls, the knowledge of non-measurable variables becomes unnecessary (by avoiding compensation terms and regulating measurable variables only). However, these controller controls have to offer great interest frequency gains and are sometimes complex to calculate to ensure robustness and disturbance rejection. For this, see the deduced controllers of the H_∞ theory, usually very high, requiring tough simplifications on one hand and the reaction for an increased simplicity on the other hand, which led to "fuzzy controllers" which can easily be determined only in theory. Here appears once again the compromise that engineers are familiar with: "enough for, not too much for fear of"—correctors complex enough to give sufficient performance, not too complex to be effectively implementable. The engineer, with his art, has been highly praised since his/her intuitions enabled him/her to solve compromises satisfactorily, in his/her own context. This came from a long interiorized experience that made him/her an expert who often forgot where his/her knowledge came from. This context does not necessarily correspond to the context of another designer, hence the difficulty to transmit experience.

Two generations of vector control to be harmonized: Estimators and observers

It deals with the induction machine control using classical vector control, which requires knowledge of the non-measurable (in the general case) rotor fluxes which are necessary to determine the frame (precisely the rotor flux frame) with much simpler equations. They enable decoupled control which we need. In addition to that, the model is a nonlinear one—the torque is given by products and the orientation angle by a division. There must not be a unique way to solve this issue, contrary to what would be the case with a linear system. The modeling itself of a well-known machine requires hypotheses: hypotheses with or without a priori on the flux orientation? These "details" (?) clearly emphasize the concrete experience the authors have with these issues. The classical structures are hardly robust and some advanced and very competitive techniques—the Kalman filter expanded to the parameters to reach robustness—are highly appealing yet remain very requiring to be implemented. The authors themselves remain in a context where the implementation will have to be reached with simple means. This requires a "good modeling", not in the physical sense (it is known) but considering the control to design and implement. From that point of view, this modeling technique would suit many users looking for a compromise between simplicity and performance. The goal must be anticipated. For this, many studies made by the whole community and authors in particular and a sorting between sterile developments and developments that led to an efficient strategy were actually needed. Young engineers and researchers therefore have a high interest in benefiting from the experiences of their elders. They do not have time to lose on countless and dead-end paths.

Considering this point of view, some paragraphs are essential: the paragraph dealing with the "explicit technique for frame orientation", as called by the authors—this paragraph uses classical formulas of vector control—but the paragraph on "implicit technique for frame orientation" is less usual and will play a fundamental role in the future. Once again, the authors belong to the first generation of vector control, to which belonged the people who discovered the "ideal orientation", that is to say the rotor flux reference frame, which is a wonderful idea even though robustness issues derive from it.

The authors have to consider the second generation of vector controls, that uses "without a priori" models on the orientation of the reference frame in order to be less sensitive to approximations on the modeling, and that alludes to the general theory on estimators and observers.

The estimator is an open-loop process simulation. Needless to say, this simple and very natural method is hardly "robust" since not all equations and parameters of the machine are known. Yet it is a classical method. The observer consists in a closed-loop algorithm and the algorithms of the Luenberger (deterministic concept) and Kalman (statistic concept) filters have become classic today. They are subject to numerous studies in laboratories and are used in some industrial applications. These observers are more competitive yet more complex to design and heavier to implement. The theoretical performances are ideal ones if robustness is ensured, but this condition is far from evident.

This is due to a move in the context: from "control issues" themselves, the interest is now on "observation issues". The correct control strategies have been identified but such controls require non-measurable variables, since we have seen that the correct control equations highlight the rotor flux, which is a non-measurable variable. Physics allows to determine this variable through real-time calculation thanks to a wide range of equations. However, with the standard approach, these equations depend on two parameters: rotor resistance and rotor inductance. Both quantities are prone to variation, because of temperature or saturation. Stator resistance can be problematic with different perspectives. How can we make the system insensitive to variations?

We are not dealing with standard linear models. Some tricks have to be found to determine, for example, a reduced-order rotor-flux observer (this type of flux observer is easier to settle). The observer, who can also be a full-order observer, needs the implementation of a "gain matrix", which will bring about determination and robustness problems. The designer must have high-level skills which are clearly difficult to pass on.

The implementation of a Kalman filter extended to the four parameters of the model (that is to say a linearized nonlinear eighth order), with variable coefficient gains, is not a purely academic exercise, even though satisfying results can be found in scientific works and are used in sophisticated industrial applications.

Correct models, control and observation, robustness and implementation

A careful reader will very likely notice a few persistent allusions to some issues. It is all about choosing a "correct model" depending on a very strict criterion, which is not an easy task. The "correct model" will allow to solve this problem, which is not posed in explicit terms and not properly formalized. Most of the time, a clear statement of the problem is reached only when a solution has been found. The correct model should answer four questions simultaneously, intellectually apart but interrelated in practice: control, observation, implementation and robustness. The control should be "competitive" in terms of dynamics. Observation will have to make up for the instrumentation failures (since the sensors are missing). Implementation should be "simple" and the whole should be "robust". This is quite promising.

When considering control issues, the robustness concept has become highly pregnant. Linguistically speaking, a system is robust if it carries on working properly with changing conditions. It obviously brings about many questions. What does "working properly" mean? What do "changing conditions" correspond to? Practice has it. If a device is still working after several approximately ten practical sessions with imaginative students and its proportion of damage, then it is surely robust. Is it possible to sell a speed variator and wait for the users (customers that have paid for your product) to show their (dis)contentment? Of course, robustness can be observed thanks to reliable theoretical tools (structured singular value, μ-analysis). They are more commonly used to test the dynamic behavior of a system when parameters vary within physically acceptable limits. They are quite complex and do not always lead to the synthesis of robust controls. Empirical methods, often established from simulations, can also be used. The designer develops a control that is supposedly robust, and then, he carries out a parameter variation and checks the robustness of the device through simulation. This kind of an empirical practice is fairly common.

One good thing is the simplicity of this method, but it requires the combination of different elements: a reliable simulation process, and representative digital tests. Mr. Benoît Robyns and his collaborators brought to light a genuine and effective approach that uses the parametric sensitivity theory, another tool that does not enjoy sufficient consideration in our community.

One essential tool: the parametric sensitivity

The sensitivity theory should be taught as one of the fundamental science tools for engineers, just as variation calculation and optimization (optimization often occurs through variation calculation). It seldom happens and a curriculum reform should be carried out in that sense. Many engineers and researchers unconsciously carry on sensitivity studies. They simply change a parameter in the process to check whether it is still working. Theoretical tools have been developed and a fundamental book has also been published on the matter (Franck, 1978). Besides, the sensitivity method does not require very sophisticated mathematical tools. Rigor and application are the only prerequisites that will allow for the computation of the function derivatives of several variables and compound functions. The authors' credit relies on the application of this method to the analysis, criticism and optimization of the useful variable estimators in order to optimize their robustness.

Sensitivity is a simple and powerful tool, since it can quickly provide the characteristics of an algorithm answering this fundamental question: is it sensitive to parameter variations?

The use of such a tool goes a long way toward showing the originality of the authors. They decided to compare different controllers (some of them being particularly simple). They also compare simple estimators (that are obtained in without a priori and with a priori approaches), and notice that some of them have complementary characteristics. There is no ideal solution but some are better than others in specific operating situations.

As it is restricted to "specific operating situations", the best option might be to adapt the control and observation to these situations (low or high speed, small or big load torque). A constant optimization process needs to be carried out choosing the best control and observation. It appears to be an "adaptive control", but this method is not well accepted in the field of motor control where transients can be very long and very short at the same time. Identification and controller choice cannot be performed simultaneously (considerable delays), while applying the controls. In such a context, those "(pseudo) adaptive controls" are actual "nonlinear controls". An algorithm can easily compute the best estimation and best control with the help of measurable variables without any delay. The question now is to know how to bring together these laws of estimation and nonlinear controls.

Fuzzy supervisor. Desensitization. Optimization

Fuzzy logic principles, established by Zadeh, were meant to provide tools to control improperly modeled systems that you still have a good knowledge of. Their reaction to stimulation is approximately known. From such basics, control rules have been defined, and expressions like "fuzzification" or "defuzzification", appeared in the scientific literature. Obviously, this method was in complete opposition to two traditions: the tradition of systems physics specialists who were looking for a precise modeling of their devices, and the tradition of automatic systems specialists who were developing very sophisticated and competitive mathematical tools. The fuzzy control hardly ever used precise physical models or powerful mathematical methods, which is both its interest (with an increasing tool complexity, it is easy to understand that a community had a rough reaction in favor of a comeback to simplicity) and its limit (why not use a model that we know is working?). Thus, in the nineties, many conflicts concerning industrial system control blew up. Some noticeable works using the fuzzy approach have been published, although most of the works done concerning electric actuators were really disappointing. Too many articles only dealt with elementary axis controls, trying to deal with the predictable variation of the mechanical parameters. The "actual control" (torque determination) results from frame orientation and current loops that are designed with traditional controls. These "fuzzy controls" were only meant to compete with PID that regulated the speed, and comparisons were established improperly with badly conceived PID. Fuzzy logic can offer much more.

That is exactly what the authors of this book established. For one thing, the authors had an issue to settle: synthesize a vector control including both the actual control and the rotor-flux estimation trying to "desensitize", with a final view to reinforce the whole system. Moreover, they have developed a genuine approach on how to use fuzzy logic. If the authors use the sensitivity functions, their aim is to desensitize the control and make it as less sensitive as possible to parameter variations. "As less sensitive as possible" means that we are dealing with an optimization problem. This problem is quite difficult, and the authors managed to find a very genuine solution. Sensitivity studies have established that two expressions of flux estimation (explicit and implicit) are complementary. Both of them have good use restrictions ("good use" referring to the desensitization to parameters variations), but how can we take advantage of each of them? The key is to use a fuzzy supervisor that will find the happy medium between the two expressions depending on the operating point. In practice, this supervisor is settled as a simple predetermined and established nonlinear function. One may wonder what a happy medium is. How can we optimize this complex system that mixes control and estimation?

Again, fuzzy logic will be roped in. It is artfully used as a heuristic optimization method that is meant to solve this problem.

In conclusion

Readers will not only enjoy a new study on vector control but they will also be introduced to useful concepts and tools that are often disregarded: the control through model inversion; the necessary variable estimation; the entanglement of the control and the estimation of nonlinear systems; the sensitivity theory; robustness and desensitization and finally, the good use of fuzzy logic as an optimization method for nonlinear systems.

Jean-Paul Louis, Professor at the E.N.S (French higher education school) of Cachan

Francis Labrique, Professor at the Université Catholique of Louvain

Acknowledgments

ISIT (Institut Supérieur d'Interprétation et de Traduction) is a graduate school offering specialized comprehensive courses. The authors would like to thank ISIT and the CRATIL research center (ISIT center for applied research in translation, interpretation and language) for their translation of this work into English. Our particular thanks go to the students who contributed to this project:

Axelle Dallot, Charlotte Dehors, Estelle Favrais, Laurent Feutrie, Amandine Guichard, Bénédicte Héron, Charlotte Kiss, Aurélie Lemaître, Emma Lepeut, Marie Loloum, Carine Mathey, Nina Mommeja, Marine Perrier, Marie-Eléonor Ryckbosch, Thorsten Schommer and Anne-Sophie Trantoul.

The translation was revised in cooperation with the Department of Translation of the Lessius Hogeschool, as part of a research project conducted by CETRA (Center for Translation Studies). We are also very grateful to Sandrine Peraldi in charge of applied research projects at ISIT for her support.

Contents

Introduction

For the past 2 decades, squirrel-cage induction machines (i.e. asynchronous machines) have enjoyed increasing success, progressively replacing synchronous and direct current machines in a vast number of industrial applications, as well as in transportation. The latest high-speed trains are a perfect example of this tendency. The success is explained by the sturdiness of the machine—which implies lower maintenance costs—and the cost-effective design. However, another major factor of this success is the steady increase in (micro) processing power, which allows for real-time control of the induction machine. This control is a great challenge, considering the complexity of the machine—which shows in the models used in designing control strategies—and the difficulty in accurately measuring basic quantities such as magnetic flux and electromagnetic torque.

Nowadays, actuation and drive systems based on inverter-fed induction motors allow for very high static and dynamic performance because of high level motor-control strategies such as vector control.

Instantaneous torque control is a prerequisite for high dynamic motor performance. In direct current machines, instantaneous torque is the product of induction flux ϕ by the current I in the secondary coil, as these two quantities are in quadrature. If the direct-current machine is a separately excited motor, flux ϕ and current I have distinct controls. In this case, when flux ϕ is constant (as is the case in permanent magnet motors), the instantaneous torque is a direct function of I, and motor control is simple. However, in induction machines, the torque results from coupling flux and current. Both these quantities are imposed from stator voltages of the machine, which are the only input quantities, as no electrical quantities can generally be imposed on squirrel-cage rotors. The vector control makes it possible to maintain two control variables, which are kept in quadrature: the flux control and the control for the torque-generating current. The main issue in implementing a vector control system for induction machines results from the difficulties in controlling or measuring the flux.

To avoid measuring it, the flux can be estimated or indirectly imposed by means of implementation of the theoretical models. However, as such commands are very dependent on the model, they are very sensitive to model uncertainties. These uncertainties are caused by the variations of stator and rotor resistance—because of temperature and skin effect—and by magnetic saturation-induced variations in inductance. Although rotor resistance usually plays a major role in vector control, it is the hardest parameter to identify accurately, especially in cage-rotor machines. Rotor resistance may vary up to 100% due to temperature changes. Uncertainties on parameters result in errors on flux magnitude and orientation in the motor, with the following consequences:

- When the orientation error is too high, the system can become unstable;
- Additional stator power consumption is needed to generate a given torque (resulting in higher losses).

A vast number of vector control strategies rely on various forms of the induction machine model. Nevertheless, control sensitivity to parameter uncertainties varies from one strategy to the other. This book presents a method for the theoretical study of parameter uncertainties. This method, based on the steady-state equations of the machine, yields results that can be extrapolated to apply to the transient state. Moreover, some stability issues can be predicted by determining the error on flux orientation. Working with this analysis method, completed by experimental trials and a number of numerical simulations, it will be possible to:

- Compare various control strategies (though not all of them will be studied);
- Suggest solutions for improving the scaling of the controls, and therefore also improving their performance;
- Introduce new control strategies, especially multi-model controls with fuzzy logic supervisors.

The first chapter is a reminder of the main physical laws and fundamentals of electromechanical energy conversion through magnetic systems. In particular, the chapter continues steadily with a demonstration that all these magnetic phenomena are governed by the same universal principle of determinism, which corresponds to the principle of causality in physics. A basic example, based on a machine composed of a rotor coil and a stator coil, is studied so as to emphasize the articulation of the cause-effect relationships that govern any electromechanical conversion.

Chapter 2 introduces the mathematical models of the induction machine, for both wound-rotor and squirrel-cage machines. Using matrix formalism, the model of the machine is first explained in a three-phase frame of reference linked to the power supply. Mathematical transforms are then introduced and applied in order to substitute simplified values for electrical quantities, thereby simplifying both the calculations and the models. A general model of the machine is subsequently presented, along with models which are better adapted to control design.

Chapter 3 suggests a number of control strategies for squirrel-cage machines. Designing a system control device becomes more complicated as the complexity of the system itself increases. This is why we will show, in the first part of this chapter, how Causal Ordering Graphs (C.O.G.) formalism enables the systematization of the conception approach and the standardization of a control system structure. An introduction to the principles of vector control of induction machines follows. Several control strategies based on rotor-flux orientation are then described. Since these controls require an estimate of the rotor flux, different flux observers and estimators are detailed. The controls vary, based on whether or not the flux is considered oriented (control with a priori on the flux).

The subject of Chap. 4 is the theoretical analysis method for parameter sensitivity. Two variations of this method are taken into consideration, according to whether the flux is estimated with or without 'a priori'-orientation. This method is applied to various control strategies introduced in Chap. 3. The different strategies build a representative sample of the control strategies known as "direct" or "indirect"; the direct strategy includes the strategies with flux observers.

The theoretical study of parameter sensitivity suggests that combining different control strategies may help desensitize the induction machine control. The theoretical study also provides valuable information to optimize the choice of observer advantages for this desensitization. In Chap. 5, fuzzy logic methodology is applied to this new information to determine the adjustment parameters included in the combination of various controls and flux observers, depending on the operating point of the machine.

Fuzzy logic has often been used to develop fuzzy controllers to control position, speed, flux or current instead of standard controllers (P, PI or IP controllers). Some authors also use it for parameter estimation. In this book, fuzzy logic is used as a tool to choose from the combined results given by several estimators. The originality of this approach lies in the fact that the fuzzy logic supervisor is mainly developed from a theoretical sensitivity study, whereas fuzzy logic is most often used in trials on systems for which no model precise or simple enough has been developed. The objective is to keep the uncertainty factor deriving from the modeling of the motor to a minimum, while introducing fuzzy logic to manage the remaining uncertainty factor that cannot be eliminated.

Chapter 6 illustrates the benefits of the controls introduced in Chap. 5, with two applications.

The first application is inertial energy storage, integrated into an electrical network powered by a wind turbine and a diesel generator. Here, an induction machine powers a flywheel which ensures the smoothing of the total power generated by the wind, as it is irregular. This machine works either as a motor or a generator according to the need, and is therefore almost always in transient state. A multi-model control leverages the storage system dynamics.

The second application consists in a torque control system for induction machines, based on a reduced-order observer optimized through fuzzy logic. This control, designed particularly for electrical traction, is an attempt at minimizing the error on torque estimation due to parametric uncertainties, especially for the most critical operating point: the starting up of the machine, with null speed and maximum torque.

The book concludes with a comparative synthesis of the various vector control strategies exposed and with a discussion on their practical applications.

Chapter 1
Concepts for Electromechanical Conversion

This chapter is a reminder of the main physical laws and fundamentals of electromechanical energy conversion through magnetic devices. It seemed essential to reposition the modeling approach and the resulting control problems within the context of applied physics. The great discoveries of the past two centuries in electromagnetism allowed emphasizing most of the phenomena at the origin of many electrical machine inventions. It is also gradually explained that all these energetic phenomena are governed by the same universal principle of determinism, which refers to the principle of causality in physics. A basic example, based on a machine composed of a rotor coil and a stator coil, is studied so as to emphasize the articulation of the cause–effect relationships that govern any electromechanical conversion.

1.1 Definitions

1.1.1 Electromechanical Devices

An electromechanical device comprises interconnected objects aimed at performing a specific function. For instance, an accumulator battery associated with a DC (direct current) machine through a rheostat forms a speed variation electromechanical device. The machine, thus supplied, is able to move a mechanical load.

It is noticed that these devices depend on assembling mechanical and electrical objects through an electromechanical converter, the machine.

The supply is the decisive factor for this machine to work as a motor. At this point, it is possible to talk about *cause and effect*. In such a case, it refers to a macroscopic viewpoint such as "the machine produces mechanical energy because it is supplied". This statement seems trivial, but the interpretation of the causalities at each object level is the core of the modeling approach that will be dealt with at a later stage. More exactly, it is shown that a coil results from the application of a

B. Robyns et al., *Vector Control of Induction Machines*, Power Systems,
DOI: 10.1007/978-0-85729-901-7_1, © Springer-Verlag London 2012

voltage across its terminals. Thus, it is not feasible to force the current *directly* into the coil, as its own causality is inexorably oriented from voltage to current. Indeed, current cannot suffer discontinuity, as it represents the energy found in the object. In general, a device is an energy conversion actuator oriented in the causal sense, but not necessarily in the functional sense (energy transfer).

1.1.2 Energy

An object is supplied with energy if it is able to act on other objects. In the previous example, the energy accumulated in the battery causes the machine shaft rotation. Thus, in general, the specific electromechanical device converts the chemical energy into mechanical energy. This conversion is directly reversible, but the power direction does not change the device causality. The example of the simple coil proves that this statement is becoming widespread: if the sign of the voltage on its terminals changes abruptly, the sign of the power also changes with no intrinsic causality modification.

1.1.2.1 General Characteristics of Energy

Energy is of two natures:

- Potential nature due to the natural state of the specific device at rest. A loaded battery is a potential supply of electrical energy accumulated in chemical form.
- Kinetic nature induced by the movement of all or part of the device depending on its environment. The rotating shaft is a kinetic supply in mechanical form.

A remote device tends to run out of energy: the battery discharges when supplying the machine, as the machine is supposed to operate in motor mode. There are often key intermediary accumulations in energy conversion, but there is always a degradation of part of the energy in heat form. Physics laws (Joule effect, frictions) legitimate this practical observation. Contrary to accumulations, losses contribute to the device stability.

1.1.2.2 Expression of Energy

Whatever its nature is, the $\mathrm{d}W$ variation of the energy found in an object or device always results from the product of two factors:

$$\mathrm{d}W = F_c \, \mathrm{d}F_q \tag{1.1}$$

F_c : a characteristic factor,
$\mathrm{d}F_q$: the variation of a quantity factor

Table 1.1 Energy characterizations in electromechanical devices

Nature of energy	Characteristic factor	Quantity factor
Kinetic mechanical	Speed	Movement quantity
Potential mechanical	Force (or torque)	Position
Kinetic electrical	Current	Flux
Potential electrical	Voltage	Quantity of electricity

The characteristic factor allows the external observation of the device development. The quantity factor determines the available energy volume, and therefore the development possibilities. Table 1.1 defines factors for standard electromechanical devices.

1.1.2.3 Power and Energy

The power p (expressed in **watts**, symbol W) is the product of a specific **potential** factor F_{cP} by a specific **kinetic** factor F_{cC}. The energy W (expressed in **joules,** symbol J) is the sum of this power, represented by the integration:

$$p = F_{cP}\, F_{cC}, \qquad W = \int_t p \,\mathrm{d}t \tag{1.2}$$

The power measures the energetic flux between the objects of a device or between the devices themselves. Any power conversion always means changing the quantity or characterization of the factors.

1.2 Electromagnetism

The laws of electromagnetism constitute the very concepts of electromechanical conversion. It was in fact the French physicist André-Marie Ampère (1775–1836) who gave the interpretation of the phenomenon observed by Danish physicist Hans Christian Ørsted (1777–1851), which meant the deviation of a magnetic needle located next to a straight conductor and run through by current.

1.2.1 Ampère's Law

Ampère put forward various hypotheses based on experience and intuition: for instance, the existence of particle currents in an armature core, which not only run through in parallel planes, but also in normal planes, according to the axis of the armature core. Furthermore, as currents run through two straight conductors, forces tend to orient these conductors in the same way and direction.

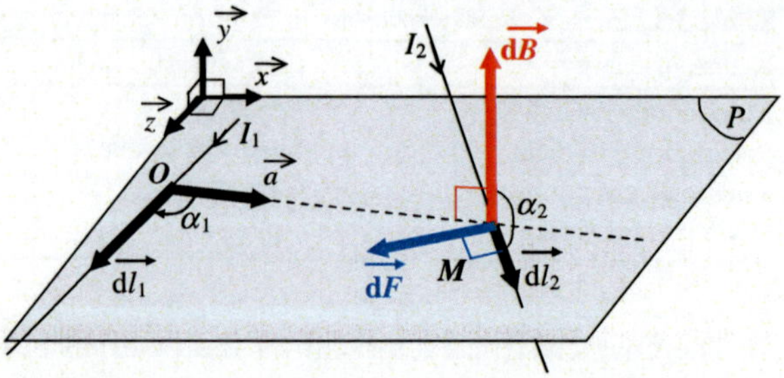

Fig. 1.1 Ampère's force—interaction between two currents

In this way, two portions of conductors $\mathrm{d}l_1$ and $\mathrm{d}l_2$ located in the same plane (P) are studied (see Fig. 1.1). These conductors, at a distance from the length OM, are respectively run through by currents I_1 and I_2.

The vector $\vec{a}$ is a unitary vector of origin O and is oriented towards the point M. The plane P is defined by vectors $\vec{\mathrm{d}l_1}$and $\vec{a}$ which both define the angle α_1. According to Ampère, a force $\mathrm{d}F$ is applied on the conductor portion $\mathrm{d}l_2$. This force, normal to $\mathrm{d}l_2$ and located in the plane (P), has a direction that tends to superpose the elements of currents $I_1\,\overrightarrow{\mathrm{d}l_1}$ and $I_2\,\overrightarrow{\mathrm{d}l_2}$. When writing α_2 the angle set by $\mathrm{d}l_2$ with the normal to the plane (P), Ampère's formula results in the following force module:

$$\boxed{\mathrm{d}F = \frac{\mu_0\mu_r}{4\pi}\,I_1 I_2 \frac{\mathrm{d}l_1\ \mathrm{d}l_2\ \ \sin(\alpha_1)\ \sin(\alpha_2)}{OM^2}} \tag{1.3}$$

μ_0 refers to the air magnetic permeability (equal to $4\pi\,10^{-7}$ Tm/A) and μ_r to the relative permeability of the material in which the conductors are found. When it is air or vacuum, then $\mu_r = 1$, otherwise the value of μ_r can be expressed in several hundreds, even thousands of units.

1.2.2 Laplace's Law

The following proposition comes from Ampère's formula: at point M, a vector quantity—characteristic of the space at this point—is able to determine a force $\mathrm{d}F$ applied on the element of current $I_1\mathrm{d}l_1$. This vector quantity, called **magnetic induction**, results from Ampère's formula:

$$\mathrm{d}B = \frac{\mu_0\mu_r}{4\pi} I_2 \frac{\mathrm{d}l_2 \sin(\alpha_2)}{OM^2} \tag{1.4}$$

It is generally written as B, if the induction received at one point of an electrical conductor depends on several other electrical circuits, or if it is due to the effect of an armature core. French physicist, mathematician, and astronomer Pierre-Simon, Marquis de Laplace (1749–1827) used Ampère's result in this way:

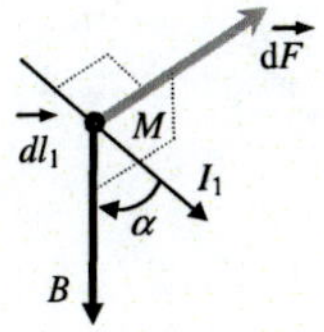

Fig. 1.2 Laplace's force

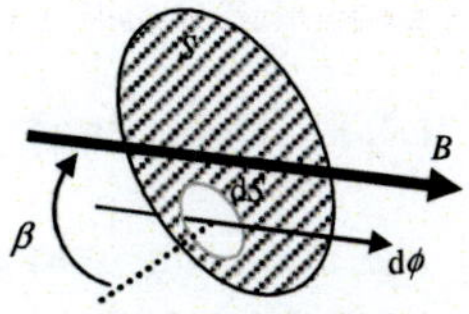

Fig. 1.3 Flux obtained by induction

$$\boxed{dF = B\, I_1\, dl_1\, \sin(\alpha)} \tag{1.5}$$

This force dF is normal to both the conductor and the magnetic induction (see Fig. 1.2). If the current or induction changes its direction, the force also changes its direction. This is also expressed by the vector product:

$$d\vec{F} = I_1 d\vec{l}_1 \, \Lambda \vec{B} \tag{1.6}$$

In practice, the force direction can be explained by the corkscrew rule turning from current to induction, or by the right-hand rule that makes a trirectangular trihedron: thumb *current*, forefinger *magnetic field*, middle finger *force*.

1.2.3 Induction Flow

Induction flow is the surface induction that runs through a material, which is subject to a magnetic induction B (see Fig. 1.3). The surface element dS located in the inclined plane set at an angle β according to the induction vector is studied. It is run through by the flux $d\phi$ such as:

$$d\phi = B\, dS\, \cos(\beta) \tag{1.7}$$

If the induction has the same direction and module on the whole surface S, then:

$$\boxed{\phi = \int_S B\, \cos(\beta)\, dS = B\, S\, \cos(\beta)} \tag{1.8}$$

Like a current running through a conductor or a liquid flow rate in a pipe of variable section, the flow is conservative. It does not change along a closed outline.

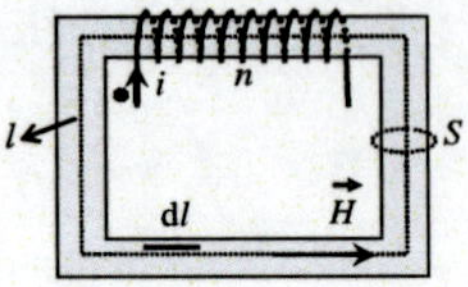

Fig. 1.4 Circulation of a magnetic field

1.2.4 Magnetic Field

1.2.4.1 Definition

The magnetic field H is a vector dimension that allows taking into account the magnetic medium, especially when referring to material other than air. As the magnetic field and the induction are oriented in the same way, both magnitudes are linked by the scalar relation:

$$H = \frac{B}{\mu} \text{ with } \mu = \mu_0 \mu_r \tag{1.9}$$

The definitions of μ_0 and μ_r are mentioned in Sect. 1.2.1.

1.2.4.2 Ampère's Theorem

Ampère's theorem expresses the relation between the magnetic field H circulating along a closed magnetic outline of length l and the section S, intertwining n times a conductor through which an electrical current i runs (see Fig. 1.4).

In other words, a winding of n, through which current i runs, is at the origin of the circulation of the magnetic field H, such as:

$$\boxed{\int_l \vec{H}\,\vec{\mathrm{d}l} = n\,i} \tag{1.10}$$

At any point of the magnetic field, if it is (rightly) assumed that the vectors $\vec{H}$ and are collinear, Ampère's theorem becomes:

$$H\,l = n\,i \tag{1.11}$$

1.2.4.3 Hopkinson's Law

The magnetic field of Fig. 1.4 is studied. From the relation (1.10), in which (1.9) and (1.8) are introduced by writing $\beta = 0$ (induction normal to the surface of n loops), may be deduced:

$$\int_l \frac{\phi}{\mu S} \mathrm{d}l = ni \tag{1.12}$$

As the flow is conservative:

$$\phi \int_l \frac{1}{\mu S} \mathrm{d}l = ni \text{ or } \phi \Re = ni \tag{1.13}$$

This relation refers to Hopkinson's law, in which $\Re$ represents the **reluctance** of the magnetic field. If section S is constant and the permeability invariant on the outline (homogeneous material), the circuit reluctance is expressed as follows:

$$\Re = \frac{l}{\mu_0 \mu_r S} \tag{1.14}$$

The products $\Re\phi$ and $n\ i$ are sometimes called **magnetomotive force.**

1.2.4.4 Inductance

In the expression of Hopkinson's law (1.13), ϕ represents the flow running through every loop of the winding. Consequently, the winding embraces a total flux equal to $n\,\phi$, which is called inductance. It defines the relationship between this total flux and current. From (1.13), results:

$$\boxed{\Phi = n\,\phi = \frac{n^2}{\Re}\, i \text{ ou } \Phi = L\,i} \tag{1.15}$$

L: coefficient called coil inductance (in **henrys,** symbol H).

Sign convention

Figure 1.4 shows the point that designs the winding entrance. An input current, positive at this entrance, creates a positive flux in the magnetic circuit.

Mutual inductance

The mutual inductance links two electrically remote windings by means of all or part of the flux running through each of the windings. Details on that topic can be found in Chap. 7 of this bibliographic reference (Hautier and Caron 1999).

1.2.5 Magnetic Material

Here, a conventional presentation on magnetic materials is given as they represent the main constituents of electrical machines. Each electrical conductor can define a magnetic circuit in the air, but an electro-magnetic circuit is essentially

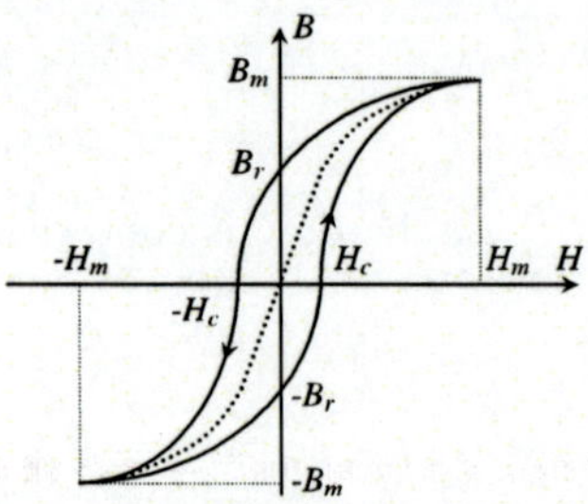

Fig. 1.5 Characteristic of a magnetic field

characterized by an electrical winding associated with a ferromagnetic material, this material having specific properties useful to the application. Usually, this material is made of a component alloy, such as silicon steel and soft iron powder composite. Without going into details, Fig. 1.5 shows—for a specific electromagnetic circuit of specific material—characteristic $B = f(H)$, obtained with a variable field the amplitude of which equals H_m.

Several phenomena that have been discovered lead to the nonlinearity of the electromagnetic circuits:

- The particular movement of point $B(H)$on a line called *hysteresis loop*: it is obtained by making the field vary from $-H_m$to H_m by acting on the current.
- The irreversibility of magnetization: the cancellation of the current—that is the cancellation of the field—leaves a permanent induction $B_r$called *remanent flux density*. The induction cancels to the field value H_c, called *coercitive field*.
- The presence of a magnetic saturation: the increase of the field leads to a decreasing variation of induction.
- The non-bijectivity of the phenomenon: to every H_m value of a maximum field corresponds a specific hysteresis loop. The highest peak in loops is called the *magnetization characteristic* (dotted line).

For every point of the characteristic $B = f(H)$, the relative permeability μ_r is determined; therefore it depends on the field, on the material and on the geometric characteristics of the magnetic circuit. It can take several thousands of unities with soft iron. Consequently, around the coordinate point (B_0, H_0) that belongs to the loop, the formula is:

$$\mu_r = \frac{B_0}{\mu_0 H_0} \tag{1.16}$$

Simplified modeling

In the applications described in this book, since the experiment shows that, at first, the introduced complexity is useless, established models do not take into account the hysteresis phenomenon. However, the magnetic saturation phenomenon sometimes remains unavoidable and it is very often associated with a behavior model that represents only the magnetization characteristic (dotted line on Fig. 1.5)

A possible solution is to choose the following function (Khater et al. 1987):

$$\mu_r = \mu_{r0} \frac{1}{1 + \left(\frac{1-\delta}{\delta}\right)\left(\frac{B}{B_m}\right)^{2\lambda}} \quad \text{along with } \lambda \in \mathrm{N} \text{ and } -B_m < B < B_m \tag{1.17}$$

μ_{r0} is the permeability measured around the null value of induction (slope at the origin). The parameter $\delta$is the reduction coefficient of the permeability measured by the magnetic characteristic slope around the maximum induction B_m. The value of λ is then chosen to set the model characteristic on the one obtained in practice.

1.2.6 *Faraday's Law*

The British physicist Michael Faraday (1791–1867) discovered the electromagnetic induction phenomenon in 1831. The most famous experiments are those with two concentric coils that are magnetically coupled but electrically isolated; one of them is connected to a galvanometer, whereas in the other the current is modified with a rheostat. Complementary experiments consist in inserting an iron core inside both the coils or moving them away from each other but without changing the current value, and in getting a magnet near the coil that is connected to the galvanometer. In all cases, Faraday noted that each mechanical movement (such as using the rheostat, moving the coils, introducing a core, putting a magnet nearby) was the cause of a deviation of the galvanometer pointer. The presence of a current means the creation of an electromotive force (e.m.f.) in the circuit: this is called induction e.m.f.

1.2.6.1 Expression

In 1914, the physicist André Blondel (1863–1938) made an experiment that allowed him to reveal the notion of **cut flow.**

First, let us look at the system shown in Fig. 1.6, where a conductor (C) of a length L is set perpendicular in a uniform induction field B, through which a current runs. This current is considered positive thanks to a **positive** external voltage source u.

According to Laplace's law (1.5), the conductor is submitted to a force f assumed to be **positive**, which causes for a time dt a movement dx that is also positively oriented:

$$f = B\,l\,i, \quad v = \frac{\mathrm{d}x}{\mathrm{d}t} \tag{1.18}$$

where v designates the conductor's speed of movement on the rails.

It is assumed that this movement leads to an insignificant variation of resistance versus the total resistance of the circuit, which implies that the current stays constant. The power balance is then written as follows:

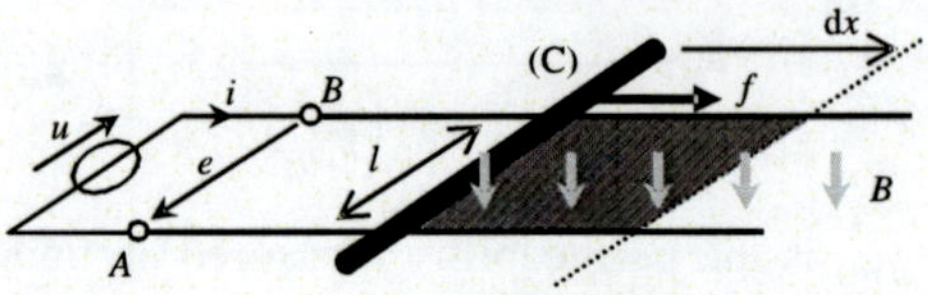

Fig. 1.6 Motor operating

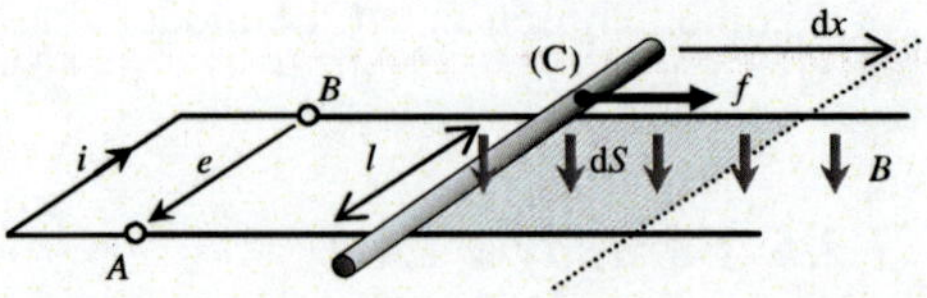

Fig. 1.7 The cut flow experiment (generator operating)

$$u\,i = R\,i^2 + f\,v \tag{1.19}$$

R is the total resistance of the circuit which is assumed to be invariable with the position of the conductor (C). Product $f\,v$ stands for the developed mechanical power corresponding to an electrical power, since:

$$f\,v = B\,l\,i\,\frac{\mathrm{d}x}{\mathrm{d}t} \tag{1.20}$$

Using the expression of the flow (1.8), then:

$$Bl\mathrm{d}x = B\mathrm{d}S = \mathrm{d}\phi \text{ and } fv = i\frac{\mathrm{d}\phi}{\mathrm{d}t} \tag{1.21}$$

where $\mathrm{d}S$ and $\mathrm{d}\phi$ respectively stand for the created surface and the flow cut by the conductor during the $\mathrm{d}x$ movement.

Now let us look at the same system, without external source ($u = 0$),in which the conductor (C) undergoes the movement $\mathrm{d}x$ because of an external mechanical action (see Fig. 1.7).

The energy conservation law implies that the mechanical power supplied in this way disappears into the Joule effect in the circuit resistance. The following formula is deduced from the power balance (1.19):

$$f = -\frac{Ri^2}{v} \tag{1.22}$$

This expression means that, since the Laplace's force is negative, the current must have changed its sign. It is opposed to the movement, which is confirmed by Lenz's Law, which enables to anticipate the induction effects as being the interpretation of the inertial effect in the system thus formed: *The direction of any magnetic induction effect is such as to oppose the cause of the effect* (Heinrich Lenz, Baltic physicist, 1804–1865). The current, which is thus created, is induced by the apparition of an induction e.m.f. e such as:

$$e = -\frac{d\phi}{dt} = -Blv \left(v = \frac{dx}{dt} \right) \tag{1.23}$$

This sign choice is the original one as movement *dx* creates a positive variation of the cut flow when the current is negative in the circuit.

Conventions

The orientation of the corresponding internal source is always the one which would provoke the flow of a positive current in the circuit when the cut flow is also positive.

The e.m.f. sign is independent from the system operating mode, which is then:

- Motor in the first experiment with a convention receptor for the electrical circuit, the supplied mechanical power being positive.
- Generator in the second experiment, the mechanical power having become negative; the convention is the same but this power is brought by an external mechanical source.

Note

Most of the time, the conventions that are used for the study of electrical machines are *receptor on the electrical side, motor on the mechanical side*, knowing that powers are considered *positive* for those operating modes. For the experiment (Fig. 1.7), if a source was connected, it could receive energy as soon as mechanical power reverses and goes beyond the Joule loses balance.

1.2.6.2 Generalized Ohm's Law

The mesh equation applied to Fig. 1.6's system is written as follows:

$$u = Ri - e \text{ ou } u = Ri + \frac{d\phi}{dt} \tag{1.24}$$

This expression is known as the generalized Ohm's law. It appears to come from the association of the original Ohm's law and Faraday's law. This type of writing is common but the causality of physical phenomena brings us to another form (see paragraph 1.3.2, Note 1 and paragraph 1.3.3).

1.2.6.3 Cut Flow, Flux Linkage, Orientation

The previous notion of cut flow—provoked by an alteration of the electrical circuit through which a magnetic field runs—corresponds to a flow variation. Consequently, when no alteration occurs but the flow varies, it can be associated with an equivalent cut flow. In other words, the phenomena are the same, that is to say that

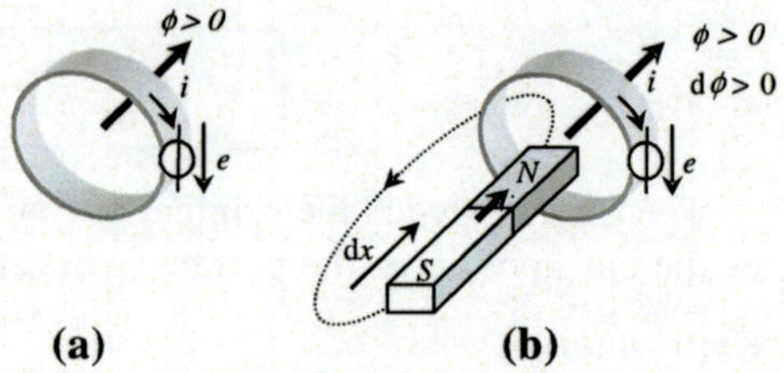

Fig. 1.8 Flux linkage variation

the principle of conservation of energy described by Lenz's law will be verified when the flow varies in a winding.

In the coil shown in Fig. 1.8a, it is assumed that the **positive** current induces a flow that is also **positive** (the orientation being given by the rule of the cork screw turning in the same direction as the current); the previous convention imposes to give to the induced e.m.f. the same direction as the current.

When a South–North oriented magnet (see Fig. 1.8 b) is near the coil, the flow will grow in the coil. Lenz's law indicates that the current has to decrease so that the e.f.m. is negative and equal to:

$$e = -\frac{\mathrm{d}\phi}{\mathrm{d}t} \tag{1.25}$$

Supposing no loss occurs, the law of conservation of energy implies that the power balance is null. The mechanical power spent on the variation of flow by moving the magnet balances the electrical power needed to prevent the flow from growing, that is:

$$p_m = -e\,i = i\frac{\mathrm{d}\phi}{\mathrm{d}t} \tag{1.26}$$

The relation between mechanical energy and electrical energy is deduced from the following expression:

$$\boxed{\mathrm{d}W_m = i\,\mathrm{d}\phi} \tag{1.27}$$

1.3 Causal Representation

1.3.1 Preamble

This section deals with the representation mode of the systems in the shape of Causal Ordering Graphs (C.O.G.). This very useful tool forces to construct the model by following the rules imposed by natural causality. These rules are specific to each component of the system (Hautier and Faucher 1997; Louis et al. 2004).

One of the main advantages of the representation is that the inverse modeling, that is to say the control structure, is easily obtained.

Observation and knowledge are the basis of informational modeling, that is to say the basis of the representation accounting for causes and effects that rule the mechanism of a system. The universal laws of physics, interpreted with stiffness and logic, are the basis of the construction of a C.O.G. model. The presentation of this formalism is now given by analyzing the basic laws ruling electromechanical conversion.

1.3.2 Causality of Faraday's Law

Let us try to understand how the e.m.f. arises and even how it works by looking at Fig. 1.1. In order to move the conductor (C), the mass of which is supposed to be equal to m, a driving effort f_e must be applied such as (a motor convention has been chosen):

$$R1 \rightarrow m\frac{\mathrm{d}v}{\mathrm{d}t} = f - f_e \rightarrow v = \frac{1}{m}\int_t (f - f_e)\,\mathrm{d}t + v_0 \tag{1.28}$$

v_0 being the initial value of the conductor's speed ($v_0 = 0$ if the conductor is fixed).

This writing recalls the motor convention. f_e must be negative, so that positive speed and moving are obtained; according to Lenz's law, f is the reactive Laplace's force, which also has to be negative.

The relation (1.23) directly gives the e.m.f. of induction according to the speed (that made it rise):

$$R2 \rightarrow e = -Blv \tag{1.29}$$

This e.m.f. is at the origin of the negative current i, which causes the reaction force (according to Lenz's law). The reaction force is opposed to the cause that created the overall phenomenon, thus the driving effort. As a consequence:

$$R3 \rightarrow i = \frac{e}{R}, \quad R4 \rightarrow f = Bli \tag{1.30}$$

Thus, the cut flow results from the mechanical action. Already expressed by (1.21), this flow variation can be expressed by integrating the e.m.f. of induction (1.23), such as:

$$R5 \rightarrow \frac{\mathrm{d}\phi}{\mathrm{d}t} = -e \quad \rightarrow \phi = \int_t -e\,\mathrm{d}t + \phi_0 \tag{1.31}$$

in which ϕ_0 is the initial value of the flow.

Table 1.2 Causal organization of Faraday law's equations

Electrical relations:	Electromechanical coupling:
$R_1 : v = \frac{1}{m}\int (f - f_e)\,\mathrm{d}t + v_0$	$R_2 : e = -Blv$
$R_3 : i = \frac{e}{R}$	$R_4 : f = Bli$
$R_5 : \phi = \int -e\mathrm{d}t + \phi_0$	

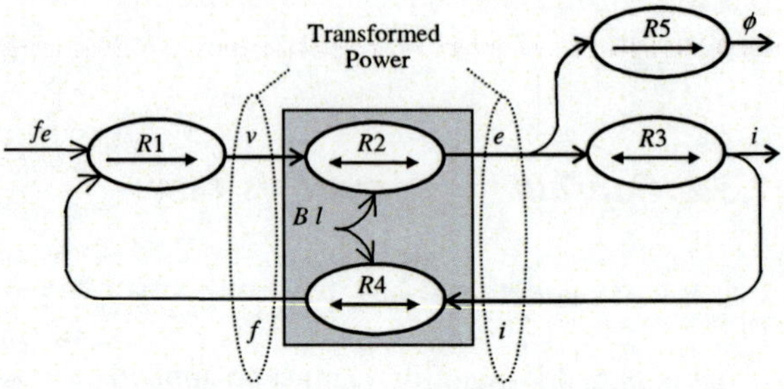

Fig. 1.9 COG model of Faraday's law

Then, in the strict perspective of phenomena related physics, it seems inconceivable to consider the e.m.f. as the flux derivative. The relations from $R1$ up to $R5$, given by the expressions (1.28)–(1.31) Table 1.2, represent the system causalities. In order to represent them, they are considered as held by an informational processing unit. The association of these processors constitutes the COG model given in Fig. 1.9.

Description

- The $R1$ relation is with **intern causality**, since it represents an accumulation of kinetic energy due to the conductor's mass (C), supposed to move without friction on the rails. This assertion is obviously a modeling hypothesis. The unidirectional arrow shows the causality direction, the speed being the consequence of the efforts, and not the inverse. The relation is said to be **causal** and cannot be directly inverted. Indeed, **since it corresponds to an accumulation of energy, the natural causality is always under integral form.**
- The $R3$ relation is with **external causality** because it represents a power dissipation in the circuit resistor. The bidirectional arrow means that the input and output variables are determined by the system topology. Thus, the appearance of the e.m.f. instantly provokes the current flow; the relation is said to be instantaneous or rigid. It can be directly inverted, meaning that the electric quantity applied to a resistance -heat sink- is either the current or the voltage, the other variable being a consequence. When serially inserted in an inductive circuit, the current is in fact imposed to the resistance (see Sect. 1.3.3); when inserted in parallel in a capacitive circuit, the voltage is imposed.
- The rigid $R2$ and $R4$ relations have been gathered in the same frame because they represent the electromechanical conversion realized by the device. They correspond neither to an accumulation nor to a power dissipation, but to **a power transformation**. They are also relations with external causality that

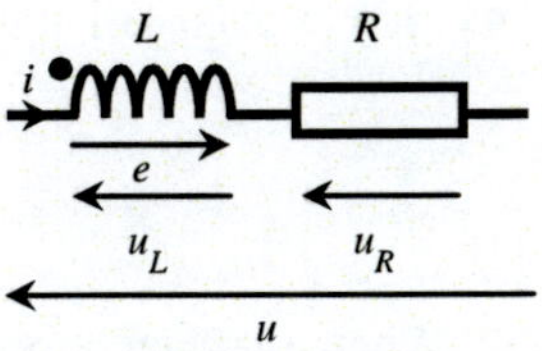

Fig. 1.10 Inductive serial circuit

cannot be found with inverted causality. Indeed, speed is necessarily an energetic quantity coming from an integration, that is to say the exit point of an accumulation processor like *R*1. As the direction of *v* is predetermined, the direction of the three other variables (*e, i, f*) obviously results from it.

- The *R*5 relation seems to be causal, expressing the cut flow via an integral **actuator**, without representing an energetic accumulation like *R*1. It is then emphasized that Faraday's e.m.f. does not result from the cut flow derivative, but that it is a rigid relation between the moving speed of the conductor and the surface induction *B S*. The writings (1.23) and (1.25) are improper uses that lead to the giving of a fallacious character to Faraday's law. It is more appropriate to express the cut flow by the e.m.f. integral and by the moving speed.

***Comment* 1**

In the writings of the relations, it is set that the first part will bear the effect of the cause(s) then written in the second part. This habit favors the physics direction that can be founded only upon natural or integral causality, or upon topological—extern—causality. Then, in order to simplify the writing, an integral causality will only be described in the shape of a differential equation respecting the arrangement thus proposed, without excluding the initial conditions, implicitly present.

***Comment* 2**

The product of the input and output variables of the energetic processors (*R*1, *R*3) gives the instantaneous power at issue in the physical object they represent. The relations of power transformation *R*2 and *R*4 are indivisible. They form what we call the **gyrator**, because each processor changes the energetic nature of the variable present at its entrance. For instance, the current corresponding to a kinetic energy is at the origin of the effort, potential variable (see Table 1.1). In accordance with the definition given in Sect. 1.1.2.3, the transformed power is given from one part to another by the product of the kinetic and potential variables, taken two by two (elliptic markers in Fig. 1.9).

1.3.3 Electromagnetic Energy

A serial circuit consisting of a coil associated with a resistance (see Fig. 1.10) is studied.

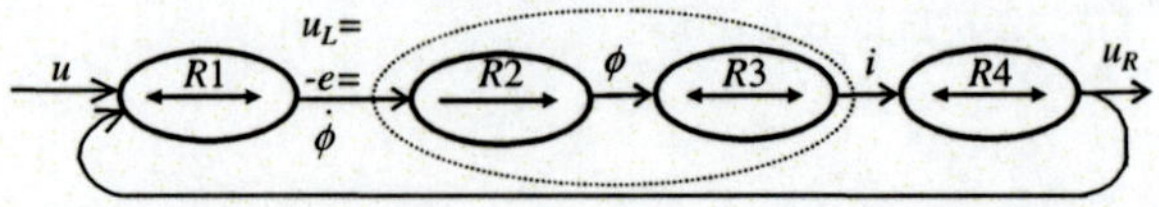

Fig. 1.11 COG model of the serial inductive dipole

When a voltage u is applied, the flux variation is given by Faraday's law, which is to be expressed in the following causal form:

$$R2 \rightarrow \phi = \phi_0 + \int_t -e\,\mathrm{d}t \quad (u_L = -e) \tag{1.32}$$

Indeed, while the voltage is applied, as the current is null, the flux derivative equals the voltage, so that the flux integrates the voltage u_L at the coil extremities. e is the auto-induction e.m.f. whose direction is given by application of the sign conventions in Sects. 1.2.4.4 and 1.2.6.1. The flux is linked to the current by the Hopkinson law, taken from the form of the expression (1.15) with the linearity hypothesis:

$$R3 \rightarrow i = \frac{1}{L}\phi \tag{1.33}$$

The current is indeed the immediate consequence of the appearance of the flux, so that the causality of the resistance is set as follows:

$$R4 \rightarrow u_R = R\,i \tag{1.34}$$

In order to determine the resulting voltage at the coil extremities, the Kirchhoff's voltage law is used:

$$u - u_R - u_L = 0 \tag{1.35}$$

Taking into account the fact that the voltage u_R is imposed by the resistance element and that the voltage u is imposed by an external generator, the electric quantity determined by the Kirchhoff's voltage law is:

$$R1 \rightarrow u_L = u - u_R = -e \tag{1.36}$$

These four relations lead to the COG model given in Fig. 1.11. The processor bearing the $R1$ relation is a mathematical operator, energetically neutral, translating the mesh law oriented according to extern causalities. The $R2$ and $R3$ processors represent the coil in which the power at issue is expressed (still with the $R3$ relation linearity hypothesis):

$$p_{\mathrm{mag}} = i\,\frac{\mathrm{d}\phi}{\mathrm{d}t} = i\,L\frac{\mathrm{d}i}{\mathrm{d}t} \tag{1.37}$$

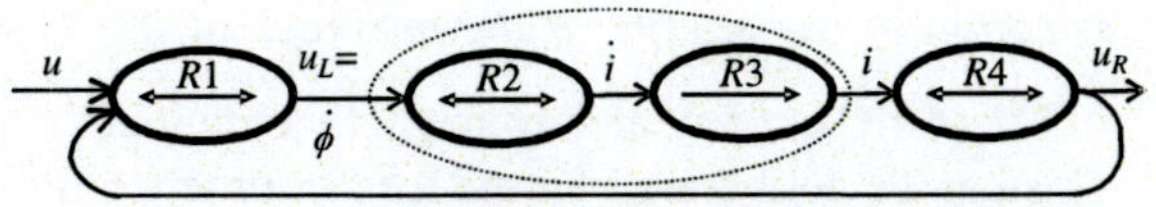

Fig. 1.12 The COG model variant of the inductive dipole

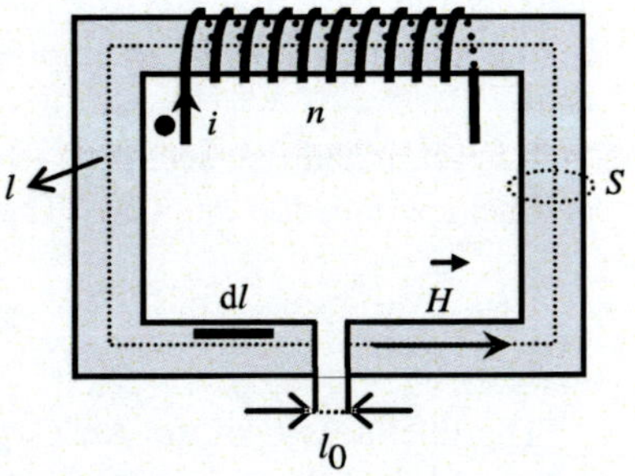

Fig. 1.13 Electromagnetic circuit with Air Gap

At a given moment, the accumulated energy under electromagnetic shape is worth:

$$\boxed{W_{\text{mag}} = \int_t p_{\text{mag}} \text{d}t = \int_i i\,L\,\text{d}i = \frac{1}{2} L i^2} \tag{1.38}$$

The current is the characteristic factor of this energy. As the energy of an isolated system cannot suffer discontinuity, *i* is the status variable of the studied dipole. The causality of a coil is internal and independent of the circuit topology where it is found. If the induction is constant (linear case), the (1.33) relation is valid between the flux and current derivatives. In those conditions, the current derivative can be shown by permuting the *R*2 and *R*3 processors functions (see Fig. 1.12). It leads to:

$$\begin{gathered} R1 \rightarrow \dot{\phi} = u - u_R,\ R2 \rightarrow \dot{i} = \frac{1}{L}\dot{\phi} \\ R3 \rightarrow \frac{\text{d}i}{\text{d}t} = \dot{i},\ R4 \rightarrow u_L = Ri \end{gathered} \tag{1.39}$$

Following those two causal representations, it must be remembered that, in an inductive circuit, the magnetic field (or its derivative) is the consequence of the induction (or of the induction derivative). This is a fundamental point for the modeling of all kinds of electric machines.

In both cases, the *R2* and *R3* processors can be fused into one for which u_L is the entrance and *i* the exit. This fusion is only possible when the electromagnetic circuit is supposed to be linear (value of the constant induction). The characteristic of the material used can be taken into account by applying the behavioral model proposed in Sect. 1.2.5, but the representation becomes more complex.

Location of the energy in the presence of Air Gap

The Air Gap of an electromagnetic circuit is the part formed by a volume of air. This can be deliberately introduced in order to limit the overloading effect. It can also appear naturally in a magnetic circuit composed of a mobile and a fixed part. This is the case for electric machines and especially for induction machines. Figure 1.13 resumes the example of Sect. 1.2.4.2, with the introduction of an Air Gap.

It is reminded that section S is constant. It is supposed that all the field lines are collinear within the Air Gap. The appliance of Hopkinson's law gives:

$$\Re = \frac{l}{\mu_0 \mu_r S} + \frac{l_0}{\mu_0 S} \tag{1.40}$$

The thickness l_o of the Air Gap is often negligible compared to the total length l. The value of the air permeability leads to a sensible increase of the reluctance of the magnetic circuit. The value of the induction is lowered as it equals:

$$L = \frac{n^2}{\Re} = \frac{n^2 S}{\frac{l}{\mu_0 \mu_r} + \frac{l_0}{\mu_0}} \tag{1.41}$$

For practical purposes, the Air Gap varies from 0.5 mm to several mm. The relative permeability varies from some hundreds to some thousand units.

If the geometrical conditions and the value of the magnetic permeability are:

$$\frac{l_0}{\mu_0} > \frac{l}{\mu_0 \mu_r} \tag{1.42}$$

Then the induction value is essentially due to the air permeability and to the thickness l_o, so that the expression (1.36) allows the affirmation that the majority of the energy is stored in the Air Gap. This result is characteristic to electric machines where, very often, the hypothesis is to assert that the electromagnetic energy is indeed concentrated in the Air Gap. The following shows that **the power exchanges between fixed and mobile parts are performed through this energy.**

Note that for the design of an electromechanical conversion device, the magnetic circuit is used both for bearing windings and as a flux concentrator. The flux concentrator is thus canalized in the Air Gap, whose weak length allows the reduction of the nonlinear effects of the material on the characteristic generally obtained. In this way, the induction values are sometimes considered invariant at first approach. Their non-stable condition due to the variable magnetic status of the iron can be taken into account thanks to a behavioral law of the type (1.17). Finally, these physical reasonings help to notice that the bigger the material relative permeability, the better the power transmission through the Air Gap.

Fig. 1.14 Rotor and stator of an induction machine

1.4 The Electromechanical Conversion

1.4.1 Principle of Electromechanical Conversion

A simple observation of any kind of electric machine emphasizes the presence of an intermediate energy. To prove this, it is sufficient to dismantle an induction machine (see Fig. 1.14). The stator (the fixed part to the right) that receives the electric power only has a physical link rotating with the rotor (the rotative part to the left), which obviously allows obtaining the mechanical energy. It is shown that the power transfer between both frameworks is performed thanks to the variation of the stored-up electromagnetic energy.

Electric energy needs a support made of an electrically conductive material, which then induces unavoidable Joule losses. The material composing the electromagnetic circuit also induces magnetic conduction losses, due to Foucault currents and hysteresis called "iron losses". Similarly, the mechanical energy transfer needs a device—stages, which allows the movement of a mobile part in relation to a fixed part, at the origin of the friction losses. This physics reasoning allows the easy prediction of the way the provided total energy will spread into such a machine. According to Ampère's theorem, Laplace's and Faraday's laws, it seems able to work either as a motor, or as a generator. In the first case, it receives electric energy and gives back mechanical energy. In the second case, the transformation is inverted. This energetic balance is analyzed, taking into account the intermediate magnetic energy needed. An electromechanical conversion structure with two electromagnetic frames (stator and rotor) is then studied. To make it work as a motor (mechanical source), the total variation of the electric energy ($\mathrm{d}W_{\mathrm{et}}$) provided to the machine is divided in:

$$\mathrm{d}W_{\mathrm{et}} = \mathrm{d}W_{\mathrm{Joule\,losses}} + \mathrm{d}W_{\mathrm{iron\,losses}} + \mathrm{d}W_{\mathrm{mag}} + \mathrm{d}W_m \tag{1.43}$$

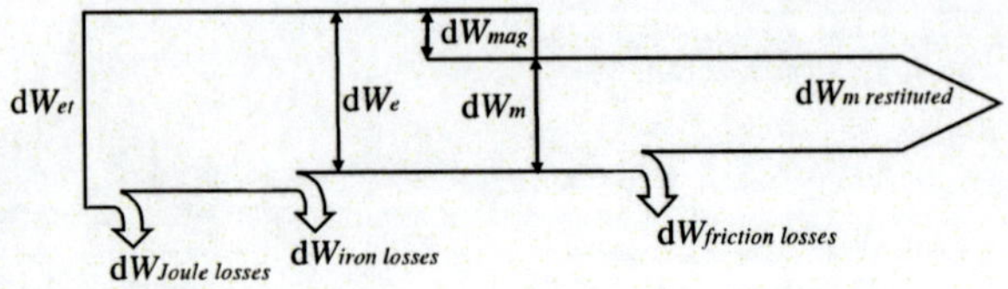

Fig. 1.15 Variational energetic balance for a motor operation

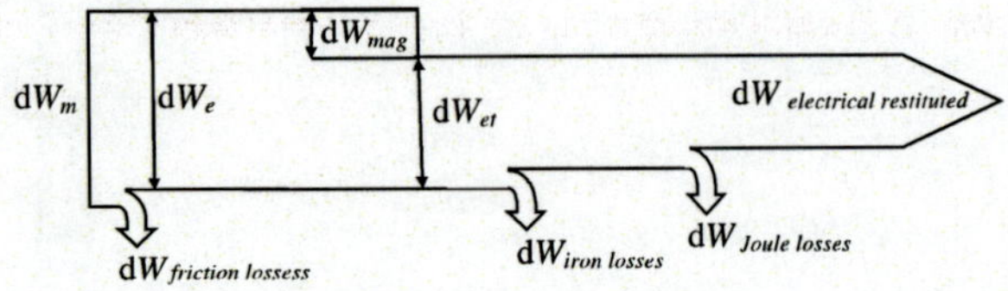

Fig. 1.16 Variational energetic balance for a generator operation

with:

$dW_{\text{Joulelosses}}$ variation of losses by Joule effect,
$dW_{\text{ironlosses}}$ variation of iron losses,
dW_{mag} variation of stocked electromagnetic energy,
dW_{m} total variation of mechanical energy

The variation of the restituted mechanical energy ($dW_{\text{m restituted}}$) is obtained from the total variation of mechanical energy by taking into account the variation of friction losses ($dW_{\text{friction losses}}$) (Fig. 1.15):

$$dW_{\text{m}} = dW_{\text{frictionlosses}} + dW_{\text{m restituted}} \tag{1.44}$$

For a generator operation, the source of energy is mechanical (Fig. 1.16) and the total variation of mechanical energy is divided into:

$$\begin{gathered} dW_{\text{m}} = dW_{\text{friction losses}} + dW_{\text{mag}} + dW_{\text{et}} \\ (dW_{\text{et}} = dW_{\text{Joule losses}} + dW_{\text{iron losses}} + dW_{\text{electrical restituted}}) \end{gathered} \tag{1.45}$$

It is assumed that each armature bears respectively n and m windings. Between them there only exists one degree of freedom in angle rotation. For a motor operation and a "receptor" convention, the generalized Ohm's law (1.24) enables to determine the electric energy absorbed by the winding j:

$$\sum_{j=1}^{n} v_j i_j dt = \sum_{j=1}^{n} R_j i_j^2 dt + \sum_{j=1}^{n} i_j d\phi_j + dW_{\text{iron losses1}} \tag{1.46}$$

In which v_j, i_j, ϕ_j, R_j respectively refer to the voltage across the bounds, the current, the total flow and the j winding resistor—cause of the energy dissipation by Joule effect of the armature 1 that can be assimilated to a stator.

The variation of the electric energy within this winding dW_{es}) can be inferred and, by extension, the one within the winding k (dW_{er}):

$$\begin{aligned} dW_{es} &= \sum_{j=1}^{n} i_j d\phi_j = \sum_{j=1}^{n} v_j i_j dt - \sum_{j=1}^{n} R_j i_j^2 dt \; - dW_{\text{iron losses1}} \\ dW_{er} &= \sum_{k=1}^{m} i_k d\phi_k = \sum_{k=1}^{m} v_k i_k dt - \sum_{k=1}^{m} R_k i_k^2 dt \; - dW_{\text{iron losses2}} \end{aligned} \tag{1.47}$$

In which v_k i_k, ϕ_k, R_k are respectively the voltage, the current, the total flow and the k winding resistor of the armature 2, which can be assimilated to a rotor.

On the one hand, variations are conserved between the net electric energy. On the other hand, there are mechanical and magnetic energy variations:

$$dW_e = dW_{es} + dW_{er} = dW_m + dW_{mag} \tag{1.48}$$

One armature is chosen as frame of reference for the angle θ. By writing T the electromagnetic torque carried out on an armature with regard to the other, the mechanical energy variation written from (1.47) is expressed:

$$\boxed{dW_m = T d\theta = \sum_{j=1}^{n} i_j d\phi_j + \sum_{k=1}^{m} i_k d\phi_k - dW_{mag}} \tag{1.49}$$

The magnetic energy is calculated either when the system is at rest, or when the angle variation equals zero. By integrating the previous equation in which $d\theta = 0$, then:

$$\boxed{W_{mag} = \left(\sum_{j=1}^{n} \int_{\phi_j} i_j d\phi_j + \sum_{k=1}^{m} \int_{\phi_k} i_k d\phi_k \right)_{d\theta=0}} \tag{1.50}$$

This expression boils down to considering the magnetic energy accumulated at a specific moment of the position of an armature with regard to the other.

1.4.2 Basic Electromechanical Conversion

1.4.2.1 Definition of the Conversion Structure

Figure 1.17 shows the front side view of two electromagnetic armatures. Each one has a winding which is coiled in slots designed for this purpose. The picture in Fig. 1.18 shows an induction machine rotor where the slots with the conductors on the inside and the coil heads on the edge can distinctly be seen. In Fig. 1.17, the winding magnetic axes are pinpointed as S and R; they carry the corresponding fluxes ϕ_s and ϕ_r whose orientations are considered to be positive when currents are also positive. The direction of the current flux in a winding is indicated by the

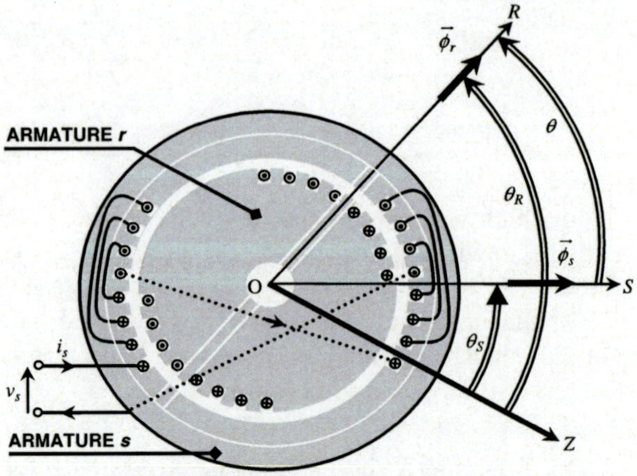

Fig. 1.17 Elementary structure with two windings

Fig. 1.18 Stator slots on an induction machine

dots (output current) and the crosses (input current) located on the edges of the conductors; the direction of the flux is defined by the corkscrew rule. Axis *Z* is arbitrarily chosen to locate the angular positions of the armatures since each of them can freely revolve around axis O. This geometry corresponds to what is known as a salient pole structure or a constant air gap structure.

Whatever the position of an armature in relation to the other, the path of the field lines of each coil is invariant (represented by a white draw for armature *r*). In standard applications, a machine magnetic circuit is made of a multi-phase winding. The field-induction relation is nonlinear and modified by two complex phenomena: saturation and hysteresis—mentioned in paragraph 1.2.5. These phenomena are reduced thanks to the air gap in which the electromagnetic energy is concentrated.

The usual hypotheses—which will be adopted later on—consists in assuming the linearity of the field-induction relation, neglecting the variable reluctance effect due to the slots, and taking into account the winding sinusoidal distribution in space around the circular armatures, as those in Figs. 1.17 and 1.18.

These hypotheses then make possible the use of a representation by inductances, in particular the mutual inductance coupling. In practice, this choice is

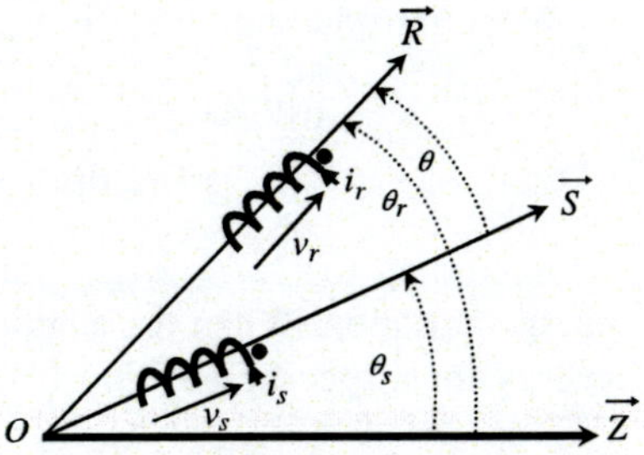

Fig. 1.19 Representation of the structure under study

almost unavoidable as first approach, bearing in mind the fact that the effect of nonlinearity caused by saturation can be taken into account by a behavior law which would then be applied to magnetic parameters.

1.4.2.2 Study of the Electromechanical Transformation

The hypotheses of linearity and of sinusoidal distribution in space allow the representation of the structure with the scheme of Fig. 1.19. We write:

L_s, L_r, M : respective inductances of coils s and r, mutual inductance between the coils when their axes coincide,

R_s, R_r : respective resistors of windings s and r,

θ_s, θ_r : respective angular positions of the magnetic axes of the coils in relation to the set arbitrary frame of reference $\vec{Z}$.

θ : relative angular position of the two coils.

The conventions are classic: a current—the input coming from the side of a winding, indicated by a dot on the scheme—creates positive flux in that same winding. According to the maximal flux principle, when the two currents represented are positive, they tend to put the coil axes in coincidence because of a torque which is also considered to be positive and which induces a rotation in a direction that is considered to be positive in motor convention.

Determination of the energies and of the torque

The clear variation in the electrical energy of the system is written as follows:

$$\mathrm{d}W_\mathrm{e} = \mathrm{d}W_\mathrm{es} + \mathrm{d}W_\mathrm{er} = i_s \mathrm{d}\phi_s + i_r \mathrm{d}\phi_r \tag{1.51}$$

where dW_{es} refers to the energy in the stator and dW_{es} to the energy in the rotor.

The flux in each coil is related to its self-induction and to mutual inductance in accordance with:

$$\begin{aligned} \phi_s &= L_s i_s + M\cos(\theta) i_r \\ \phi_r &= L_r i_r + M\cos(\theta) i_s \end{aligned} \tag{1.52}$$

When introducing (1.52) in (1.51), the electric energy variation is:

$$\begin{aligned} \mathrm{d}W_e &= i_s L_s \,\mathrm{d}i_s + M\cos(\theta) i_s \,\mathrm{d}i_r - M i_r i_s \sin(\theta)\ \mathrm{d}\theta \\ &+ i_r L_r \,\mathrm{d}i_r + M\cos(\theta) i_r \,\mathrm{d}i_s - M i_r i_s \sin(\theta)\ \mathrm{d}\theta \end{aligned} \quad (1.53)$$

According to (1.48), every variation is divided into magnetic and mechanical energy variation. When the angular variation is null ($\mathrm{d}\theta = 0$), the stored magnetic energy for a specific position ($\mathrm{d}W_{\mathrm{mag}} = \mathrm{d}W_{\mathrm{e}}$) can be deduced:

$$W_{\mathrm{mag}} = \frac{1}{2} L_s i_s^2 + \frac{1}{2} L_r i_r^2 + M i_r i_s \cos(\theta) \quad (1.54)$$

The total magnetic energy variation (when $\theta \neq 0$) is described as follows.

$$\begin{aligned} \mathrm{d}W_{\mathrm{mag}} &= i_s L_s \mathrm{d}i_s + i_r L_r \mathrm{d}i_r \\ &+ M\cos(\theta) i_s \mathrm{d}i_r + M\cos(\theta) i_r \mathrm{d}i_s - M i_r i_s \sin(\theta)\mathrm{d}\theta \end{aligned} \quad (1.55)$$

From (1.48), (1.53) and (1.55) is deduced:

$$\mathrm{d}W_{\mathrm{m}} = -M i_r i_s \ \sin(\theta)\ \mathrm{d}\theta \quad (1.56)$$

If the currents are both positive—the torque is supposed to be positive with the rotation direction during the motor operation—we can deduce that the angle variation is thus negative. This deduction is confirmed by the maximal flux principle. With Ω as the rotation speed of an armature in relation to the other one, the mechanical power is written as:

$$p_{\mathrm{m}} = \frac{\mathrm{d}W_{\mathrm{m}}}{\mathrm{d}t} = T\Omega, \Omega = -\frac{\mathrm{d}\theta}{\mathrm{d}t}, \quad (1.57)$$

The expression of the electromagnetic torque is deduced by including (1.56) in (1.57):

$$\boxed{T = M i_r i_s \sin(\theta) \text{ or } T = M\left(\vec{I_r} \Delta \vec{I_s}\right)} \quad (1.58)$$

Vector $\vec{I_r}$ thus corresponds to the current vector of i_r length, specific to armature r, which is then out of phase with angle θ (Fig. 1.19) of vector $\vec{I_s}$ of i_s length, specific to armature s.

So the effect of Ampère Forces appears again, resulting from the interaction of two currents or a field with a current.

Modeling of the structure

The generalized Ohm law applied to each winding results in:

$$\begin{aligned} \frac{\mathrm{d}\phi_s}{\mathrm{d}t} &= v_s - R_s i_s \\ \frac{\mathrm{d}\phi_r}{\mathrm{d}t} &= v_r - R_r i_r \end{aligned} \quad (1.59)$$

In (1.59), substituting fluxes by their (1.52) expressions depending on currents results in:

$$L_s \frac{\mathrm{d}i_s}{\mathrm{d}t} + M\cos(\theta)\frac{\mathrm{d}i_r}{\mathrm{d}t} + Mi_r \sin(\theta)\Omega = v_s - R_s i_s$$
$$L_r \frac{\mathrm{d}i_r}{\mathrm{d}t} + M\cos(\theta)\frac{\mathrm{d}i_s}{\mathrm{d}t} + Mi_s \sin(\theta)\Omega = v_r - R_r i_r \tag{1.60}$$

These expressions are the electric equations of the structure. The variations are divided into two parts, each of them producing induced e.m.f.:

- Induction e.m.f. are due to the moving speed of each winding in the induction field created by the other one. This is described as follows:

$$e_{\mathrm{sb}} = Mi_r \sin(\theta)\,\Omega, \quad e_{\mathrm{rb}} = Mi_s \sin(\theta)\,\Omega \tag{1.61}$$

- Transformation e.m.f., which are due to self-induction and to mutual inductance coupling:

$$\begin{bmatrix} -e_{\mathrm{st}} \\ -e_{\mathrm{rt}} \end{bmatrix} = \begin{bmatrix} \frac{\mathrm{d}\phi_{\mathrm{st}}}{\mathrm{d}t} \\ \frac{\mathrm{d}\phi_{\mathrm{rt}}}{\mathrm{d}t} \end{bmatrix} = \begin{bmatrix} L_s & M\cos(\theta) \\ M\cos(\theta) & L_r \end{bmatrix} \begin{bmatrix} \frac{\mathrm{d}i_s}{\mathrm{d}t} \\ \frac{\mathrm{d}i_r}{\mathrm{d}t} \end{bmatrix} \tag{1.62}$$

Under these conditions, the (1.59) expressions become:

$$\frac{\mathrm{d}\phi_{\mathrm{st}}}{\mathrm{d}t} = v_s - R_s i_s - e_{\mathrm{sb}}$$
$$\frac{\mathrm{d}\phi_{\mathrm{rt}}}{\mathrm{d}t} = v_r - R_r i_r - e_{\mathrm{rb}} \tag{1.63}$$

Energy balance distribution

Multiplying both parts of the (1.63) relations by the stator currents and by the rotor currents—respectively produces the balance of armature related powers:

$$i_s \frac{\mathrm{d}\phi_{\mathrm{st}}}{\mathrm{d}t} = v_s i_s - R_s i_s^2 - e_{\mathrm{sb}} i_s$$
$$i_r \frac{\mathrm{d}\phi_{\mathrm{rt}}}{\mathrm{d}t} = v_r i_r - R_r i_r^2 - e_{\mathrm{rb}} i_r \tag{1.64}$$

The global power balance is calculated by adding up the two lines and substituting the power provided by the sources in the first part:

$$v_s i_s + v_r i_r = R_s i_s^2 + R_r i_r^2 + i_s \frac{\mathrm{d}\phi_{\mathrm{st}}}{\mathrm{d}t} + i_r \frac{\mathrm{d}\phi_{\mathrm{rt}}}{\mathrm{d}t} + e_{\mathrm{sb}} i_s + e_{\mathrm{rb}} i_r \tag{1.65}$$

Developing (1.65) and taking (1.55) and (1.56) into account leads to:

$$v_s i_s + v_r i_r = R_s i_s^2 + R_r i_r^2 + \frac{\mathrm{d}W_{\mathrm{mag}}}{\mathrm{d}t} + \frac{\mathrm{d}W_{\mathrm{m}}}{\mathrm{d}t} \tag{1.66}$$

A comparison with the conceptual energy balance expressed by (1.43), shows that the electric power coming from sources (p_{et}) balances the Joule losses ($p_{\mathrm{Joulelosses}}$), the electromagnetic power (p_{mag}) and the mechanical power p_m—the iron losses are supposed to be negligible:

$$v_s i_s + v_r i_r = p_{\mathrm{et}} = p_{\mathrm{Joule\ losses}} + p_{\mathrm{mag}} + 2p_{\mathrm{m}} \tag{1.67}$$

along with:

$$2p_{\mathrm{m}} = e_{\mathrm{sb}} i_s + e_{\mathrm{rb}} i_r. \tag{1.68}$$

p_{m} stands for the mechanical power also deduced from Eqs. 1.56 and 1.56:

$$p_{\mathrm{m}} = \frac{\mathrm{d}W_{\mathrm{m}}}{\mathrm{d}t} = M i_s i_r \sin(\theta)\ \Omega \tag{1.69}$$

1.4.2.3 Causal Representation of the Elementary Electromagnetic Transformation

The separated expressions (1.64) of the energy balances underline the interactive phenomenon between both armatures. Indeed, the mechanical power appears actually on both sides; the terms $i_s(\mathrm{d}\phi_{st}/\mathrm{d}t)$ and $i_r(\mathrm{d}\phi_{rt}/\mathrm{d}t)$ correspond to the global energetic exchanges. For an observer, the mechanical power is obviously the same when focalizing on one armature or the other. This observation is due to the reciprocal action phenomena that characterize the physical systems submitted to any stress. The same effort value can be found on both sides. When a machine is running, the electromagnetic torque is created on the rotor and pushes on the stator, which induces the retaining forces on the holding means (feet or mounting flanges).

The Causal Ordering Graph (C.O.G.) of this elementary electromagnetic transformation is established from the expressions mentioned above, with regard to the modeling of the structure (see Fig. 1.20).

The symmetry of the representation is noticed in conformity with the own symmetry of this basic machine. The result, which is obtained in this way, is to be connected with the model of Fig. 1.9, provided in relation with the Faraday's law. This graph is composed of three parts:

- PART 1: Electric relations ($R1_s$, $R1_r$) are transformed by rigid processors. Each actuator integrates a heat sink processor corresponding to the Joule losses in the windings.
- PART 2: The electromagnetic relations with first the magnetic coupling relation between armatures ($R2$), which is interpreted by a parameter shifting rigid processor with angle θ; then the integration of the power derivatives ($R3_s$, $R3_r$). This set stands for the accumulated magnetic energy.

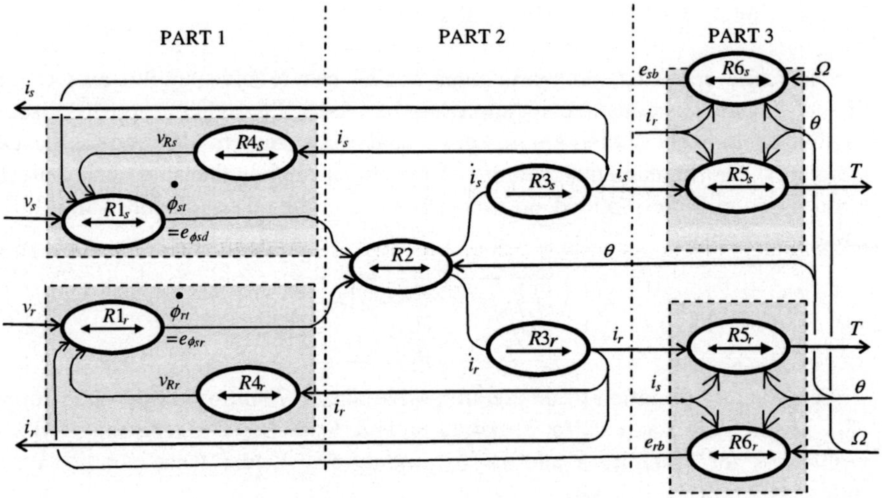

$$R1_s \rightarrow \dot{\phi}_{st} = v_s - v_{Rs} - e_{sb},\ R1_r \rightarrow \dot{\phi}_{rt} = v_r - v_{Rr} - e_{rb}$$

$$R2 \rightarrow \begin{bmatrix} \dot{i}_s \\ \dot{i}_r \end{bmatrix} = \begin{bmatrix} L_s & M\cos(\theta) \\ M\cos(\theta) & L_r \end{bmatrix}^{-1} \begin{bmatrix} \dot{\phi}_{st} \\ \dot{\phi}_{rt} \end{bmatrix}$$

$$R3_s \rightarrow\ i_s = \int_t \dot{i}_s\,\mathrm{d}t + i_{s0},\ R3_r \rightarrow\ i_r = \int_t \dot{i}_r\,\mathrm{d}t + i_{r0}$$

$$R4_s \rightarrow v_{Rs} = R_s i_s,\qquad R4_r \rightarrow v_{Rr} = R_r i_r$$

$$R5_s \rightarrow T = M i_r i_s \sin(\theta), R5_r \rightarrow T = M i_r i_s \sin(\theta)$$

$$R6_s \rightarrow e_{sb} = M i_r \sin(\theta), R6_r \rightarrow e_{rb} = M i_s \sin(\theta)$$

Fig. 1.20 C.O.G. model of the elementary electromechanical transformation

- PART 3: The electromechanical transformation relations ($R4_r$, $R4_s$, $R5_r$, $R5_s$) are supported by gyrator processors. They also represent parameter shifting rigid processors with angular position θ.

This type of model improves the understanding of these phenomena. For example, it is obvious that the torque control would require the control of currents and of the angular position. The action on currents is only possible through the voltage control and the acquirement of a constant torque, enabling to maintain the rotation. This requires an appropriate technology with a specific power supply mode.

The synchronous induction and direct current machines are electromechanical converters running because of the concepts introduced beforehand. For each of the three cases, the torque is always the result of a vector product from a "field" vector and a "current" vector; the necessity of acting on these components so as to control the torque value becomes obvious.

1.4.2.4 Modeling of the Mechanical Part

Mechanically speaking, both armatures, which can revolve around axis O (see Fig. 1.17), are characterized by moments of inertia (J_s and J_r) and by viscous friction coefficients (f_s and f_r). Indices s and r are kept to distinguish the two armatures. The fundamental relation of dynamics implies that the sum of all the torques of the forces applied on each armature is linked to the following differential equations:

$$\boxed{\begin{aligned} R7_s &\rightarrow J_s \frac{\mathrm{d}\Omega_s}{\mathrm{d}t} = T_{\mathrm{totals}} \\ R7_r &\rightarrow J_r \frac{\mathrm{d}\Omega_r}{\mathrm{d}t} = T_{\mathrm{totalr}} \end{aligned}} \tag{1.70}$$

Each one is submitted to an electromotive torque T to a resisting outer torque (T_{fix} for armature s and T_{ch} for armature r) and to a friction resisting torque—written T_{fs} for armature s and T_{fr} for armature r. Under these conditions the following results are shown:

$$\boxed{\begin{aligned} R8_s &\rightarrow T_{\mathrm{totals}} = T - T_{\mathrm{fs}} - T_{\mathrm{fix}} \\ R8_r &\rightarrow T_{\mathrm{totalr}} = T - T_{\mathrm{fr}} - T_{\mathrm{ch}} \end{aligned}} \tag{1.71}$$

along with

$$\boxed{\begin{aligned} R9_s &\rightarrow T_{\mathrm{fs}} = f_s\Omega_s \\ R9_r &\rightarrow T_{\mathrm{fr}} = f_r\Omega_r \end{aligned}} \tag{1.72}$$

The motor convention is kept for each armature so that the speeds are both considered as positive. Under these conditions:

$$\frac{\mathrm{d}\theta_s}{\mathrm{d}t} = \Omega_s, \quad \frac{\mathrm{d}\theta_r}{\mathrm{d}t} = -\Omega_r \tag{1.73}$$

The angular relation, deducted from the conventional choice of Fig. 1.17, leads to:

$$\theta = \theta_r - \theta_s \tag{1.74}$$

from which the following equation is deduced:

$$\boxed{R10 \rightarrow \Omega = \Omega_r + \Omega_s \text{ with } R11 \rightarrow \frac{\mathrm{d}\theta}{\mathrm{d}t} = -\Omega} \tag{1.75}$$

This latest result is logical since the speeds of every armature adds up. If the resisting torques and the moments of inertia were equal, the relative speed would be equal to the double of the absolute speed of each armature.

The C.O.G. model of Fig. 1.21 (a) results from Eqs. 1.70, 1.71, 1.72 and 1.75 and completes the previous graph.

Processors $R7_s$ and $R7_r$ correspond to the kinetic energies, which are accumulated in the armatures mass, and to the friction losses. This mechanical dipole (inertia friction) is similar to the inductive dipole in series studied in paragraph

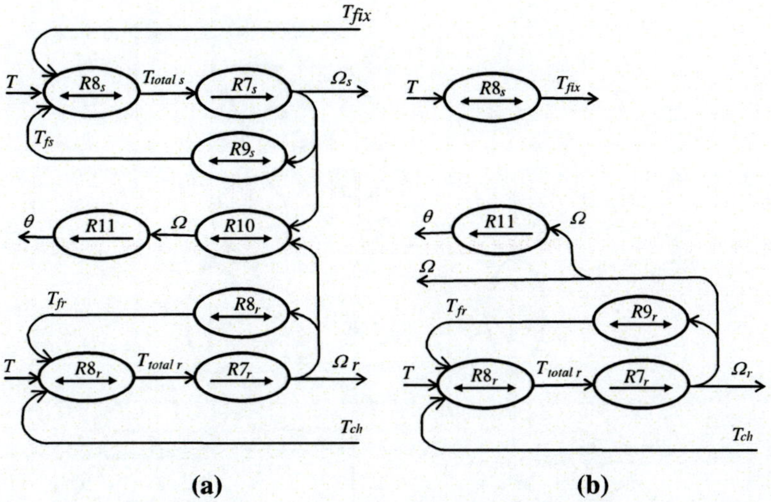

Fig. 1.21 C.O.G. of the model of the mechanical part. **a** Free armatures. **b** Fixed s armature

1.3.3. Processor $R9$ is an operator which is energetically neutral, since the exit variable is of the same kind as the entry variables. Processor $R10$ is a simple integration actuator.

When one of the armatures is kept fixed in relation to the marker, its velocity is then forced to zero. This situation corresponds to an infinitely large value of the inertia moment brought back to the corresponding armature.

If coil s is still, Eqs. 1.70, 1.71, 1.72 and 1.75 become:

$$\begin{array}{|l|}\hline R7_r \rightarrow J_r \frac{d\Omega_r}{dt} = T_{\text{totalr}}, \\ R8_r \rightarrow T_{\text{totalr}} = T - T_{\text{fr}} - T_{\text{ch}} \quad R8_s \rightarrow T = T_{\text{fix}}, \\ R9_r \rightarrow T_{\text{fr}} = f_r \Omega_r, \\ R10 \rightarrow \Omega = \Omega_r \\ \hline \end{array} \tag{1.76}$$

Figure 1.21 (b) then gives the GIC model of the two coil elementary structures. Armatures s and r then refer to the stator (which is stationary) and the rotor (which is mobile).

When associating the GIC of the mechanical part model (Fig. 1.21) to the GIC of the electromechanical transformation (Fig. 1.20), the complete model of this elementary structure is obtained. A global representation displaying the physical quantities that are exchanged between the models is illustrated in Fig. 1.22.

1.4.2.5 Simulation Analysis

The model of Fig. 1.22 is directly embedded in the Matlab simulation system software, running under the Simulink environment (MATLAB®). The model,

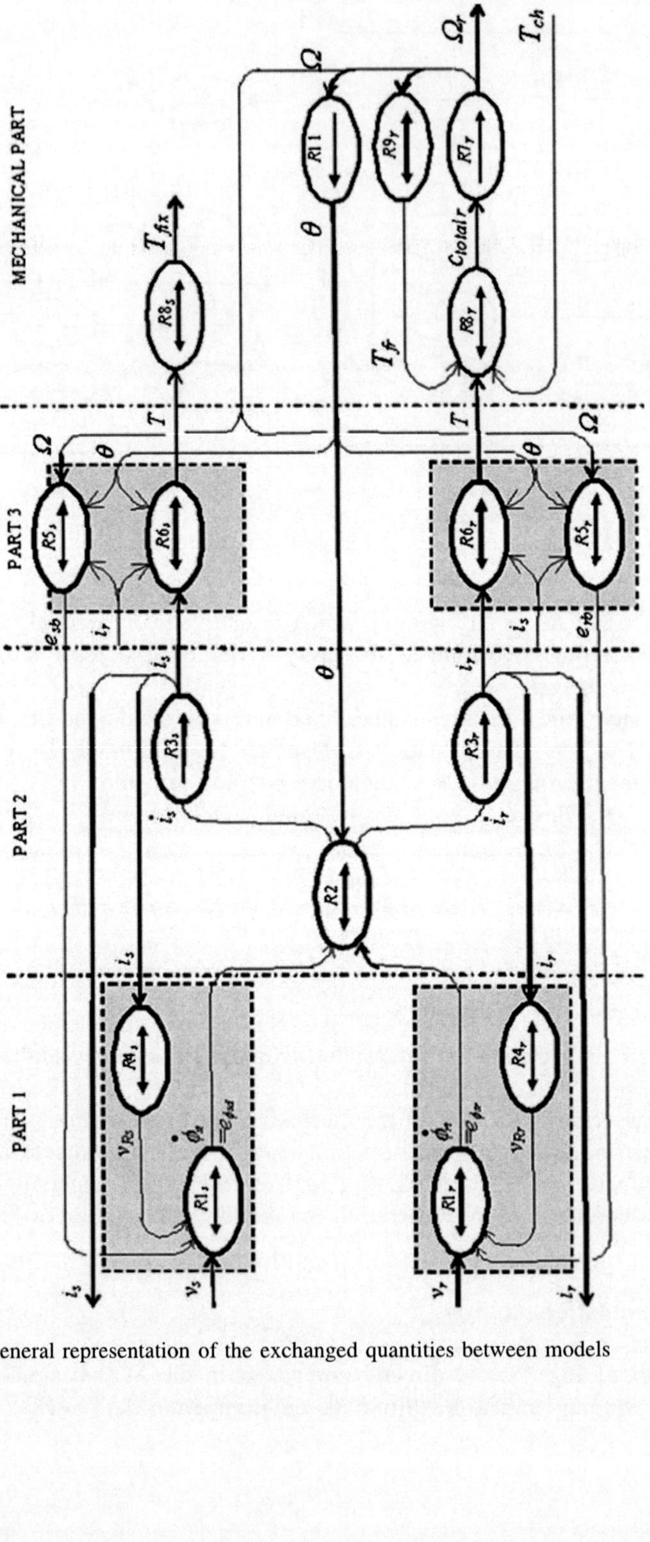

Fig. 1.22 General representation of the exchanged quantities between models

which is then obtained for simulation analysis, conforms to the Causal Ordering Graph.

Numerical values and tests conditions

- Electrical parameters:

$$L_s = 50\ \text{mH}, L_r = 20\ \text{mH}, M = 30\ \text{mH},\ R_s = 0.5\Omega,\ R_r = 0.2\Omega;$$

- Mechanical parameters:

$$J_s = 0.1\ \text{kg.m}^2,\ J_r = 0.2\ \text{kg.m}^2, f_s = 1\ \text{Nm.s/rad}, f_r = 1\ \text{Nm.s/rad}.$$

- First test (figure 1.23): currents and velocities are null, angle θ is initially shifted by $3\pi/4$ radians, both armatures can freely revolve around their common axis O. Continuous voltages are simultaneously applied to the terminals of each coil and a gravity torque is applied to armature r, that is:

$$v_s = 5\,\text{V et } v_r = 2\,\text{V}, T_{\text{ch}} = 0,2\,\text{Nm and } T_{\text{fix}} = 0$$

- Second test (Fig. 1.24): the conditions are the same as above, but armature s is stationary and the functioning is one of a machine equipped with a rotor and a stator.

Analysis of the results

For both tests, the timing evolution of the currents, the torque and the angle, the speeds and the power balances are noted. The following details from these tests are observed:

- **First case: free armatures in rotation**

The final value of the two currents is 10 A. This value is correct because after the transient operating conditions, mechanical and magnetic powers cancel each other out.

It is easily verified that generalized Ohm's laws lead to these results (see Fig. 1.23a). The final value of torque T and angle θ is not null (respectively a black line and a gray line in Fig. 1.23b). Velocity Ω (bold black line in Fig. 1.23c) is null but the velocities of each armature are opposed (Ω_s and Ω_r symbolized respectively by a discontinuous line and by a gray line in Fig. 1.23c). This can be justified for the stationary state of the mechanical part, based on Eqs. (1.70), (1.71) and (1.72):

$$T = f_s\Omega_s + T_{\text{fix}}, T = f_r\Omega_r + T_{\text{ch}} \tag{1.77}$$

With the digital values of the parameters and inputs, the following equation is found:

$$\Omega_s = 0.1\ \text{rad/s}, \Omega_r = -0.1\ \text{rad/s},$$

Fig. 1.23 Free armatures. a Currents (A). b Torque (Nm) and Angle (rad). c Speeds (rad/s). d Powers (W)

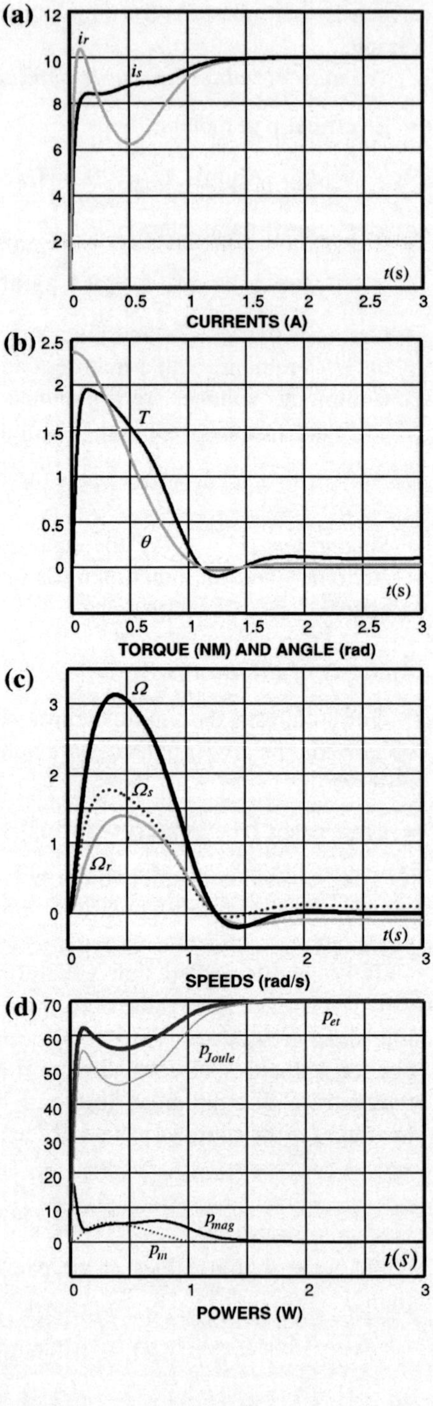

Fig. 1.24 (Stationary) stator and (free) rotor. **a** Currents (A). **b** Torque (Nm) and Angle (rad). **c** Speed (rad/s). **d** Powers (W)

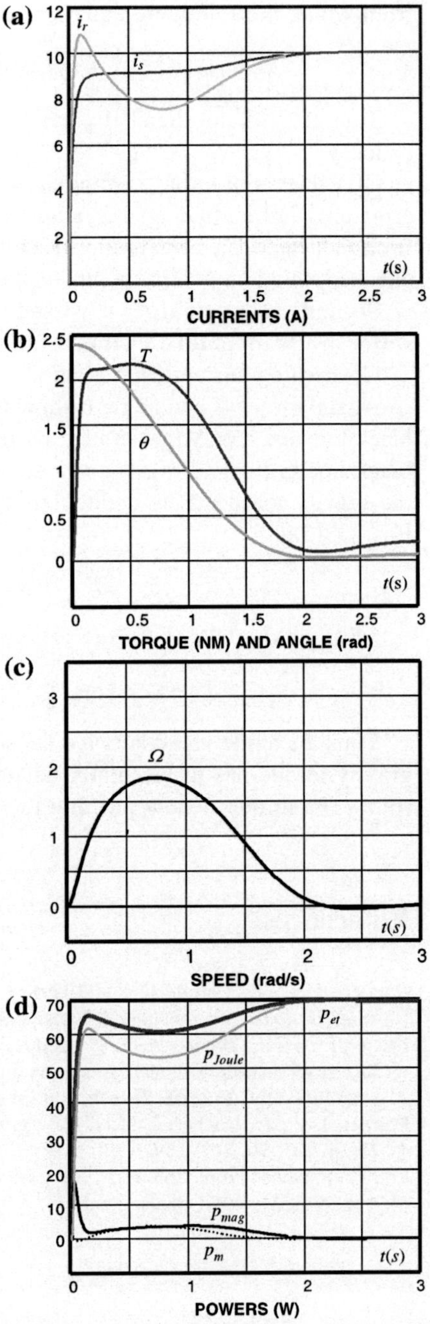

from which the following equation can be easily deduced:

$$T = 0.1\ \text{Nm},\ \theta = 33\,10^{-3}\ \text{rad}$$

Thus the two mechanical parts are revolving in the same direction [in accordance with conventions (1.73)]. The (0.2 Nm) gravity torque divides into two equal values since the coefficients of viscous friction are also equal. The two armatures are shifted by the angle value obtained above, so as to balance the friction torque on armature s. Finally, the power balance (1.67) is constantly checked, this is a crucial point for the model validation.

Second case: armature s is fixed to the frame of reference and becomes the stator while armature r stands for the rotor

Concerning currents, the results are similar to the prior cases (see Fig. 1.24a).

Again, torque T and angle θ show a final value which is not null (respectively a black line and a gray line in Fig. 1.24b). The velocity of rotor Ω is cancelled (bold black line in Fig. 1.24c). We can deduce from this that the motor torque balances the gravity torque. This deduction can be explained with the stationary state of Eqs. (1.75):

$$T = f_r\Omega + T_{ch},\quad T_{fix} = T \tag{1.78}$$

With the digital values, the following equation can be deduced:

$$T = 0.2\ \text{Nm},\ T_{fix} = 0.2\ \text{Nm},\ \theta = 66\,10^{-3}\ \text{rad}$$

Thus the angle value is twice the previous one since armatures are still and the gravity torque has to be totally balanced by the motor torque transmitted by the rotor. The torque is now pushing the stator.

References

Hautier, J. P., & Caron, J. P., (1999). *Convertisseurs statiques: Méthodologie causale de modélisation et de commande*. Paris: Editions Technip ISBN 2-7108-0745-9

Khater, F. M. H., Lorentz, R. D., Novotny, D. W., & Tang, K., (1987, March/April). Selection of flux level in field-oriented induction machine controllers with consideration of magnetic saturation effects. *IEEE Transaction on Industry Applications*, *IA-23*(2).

Hautier, J. P., & Faucher, J. (1997). Le graphe informationnel causal. *Bulletin de l'Union des Physiciens, 90*, 167–189.

Louis J. P. (sous la direction de), (2004), Modèles pour la commande des actionneurs électriques, traité EGEM, Série Génie Electrique, Chapitre 1, Hermès- Lavoisier, ISBN 2-7462-0917-9.

Chapter 2
Dynamic Modeling of Induction Machines

The cage rotor induction machine is widely used in industrial applications. Indeed, due to its conception, it has quite a low cost compared to the cost of other machines. It has a great electromechanical sturdiness and there is a good standardization between the diverse producers. Nevertheless, the relative simplicity of conception of the machine hides quite a great functional complexity, as soon as it is aimed at controlling the performed electromechanical conversion. The goal of this chapter is to establish various mathematical models, which allow understanding the machine's functioning, so as to establish its control in the next chapters.

In the previous chapter, the fundamental concepts of the electromechanical conversion have been recalled and applied to the modeling of an elementary machine, composed of a stator coil and a rotor coil. In this chapter, the physical principles are extended to the study of a three-phase induction motor. The mathematical models are established, regardless of the type of rotor technology (machines with coiled rotor that can be connected to an extern supply or squirrel-cage machines whose rotor circuit is short-circuited).

First the model of this machine is clarified in the three-axe frame linked to its supply, making the most of the matrix formalism. Then, several mathematical transformations are presented and used so as to substitute components from electrical quantity. The components will make the calculations easier and will simplify the representations. A general modeling of this machine is then presented along with some models, which are more suitable to the control design.

2.1 Presentation of the Three-Phase Induction Machine

2.1.1 Constitution and Structure

The stator of an induction machine is composed of three windings, coupled in star or in triangle. The rotor of the machine will be considered to support a winding, similar to the one of the stator, that is to say a three-phase winding with the same

B. Robyns et al., *Vector Control of Induction Machines*, Power Systems,
DOI: 10.1007/978-0-85729-901-7_2, © Springer-Verlag London 2012

Fig. 2.1 Effect of the magnetic field's move on short-circuited conductors

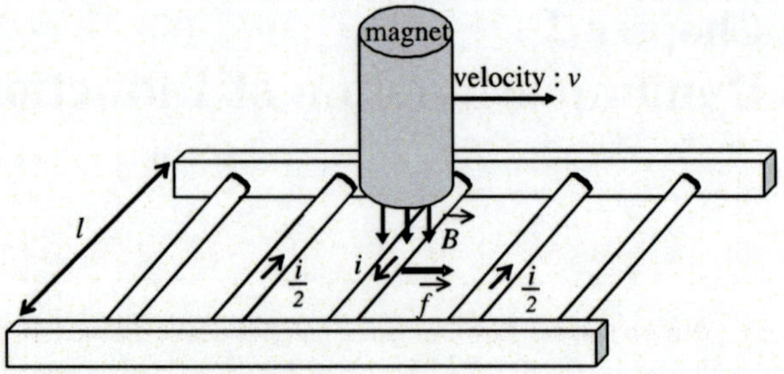

amount of poles as in the stator. These three windings are coupled in star and two machine technologies can be distinguished according to this rotor constitution.

For the first technology, the coiled rotor includes a three-phase winding, similar to the one of the stator connected in star and for which each winding free bound is linked to a rotating ring by the rotor tree. Three brushes establish contacts on this ring and allow an external connection to the rotor. This connection can be a short-circuit or a link to resistors—so that it modifies the machine's characteristics in some of the functioning areas. This connection can also be an extern supply that allows controlling the rotor quantities. This second technology is still known as "Doubly-Fed Induction Machine".

For the second technology, the rotor windings are short-circuited on themselves. The rotor construction is then broadly simplified. Industrial realizations generally use a rotor, which is composed of conducting copper or aluminum bars that are short-circuited by a conductor ring at each bound. The result appears to be a squirrel cage, which explains the current name of "cage machine". This cage rotor works the same way as a coiled rotor.

The model developed in this chapter works for both technologies, depending on the voltages across the winding terminals. If they are null, it is a cage machine, and if they are imposed by an extern supply, it is a doubly feed induction machine.

2.1.2 Working Principle

In order to understand the functioning of the induction machine, the experiment of the cut flow in generator functioning (Chap. 1, paragraph 1.2.6.1) is modified as follows. This time, the l length conductors are fixed between themselves and short-circuited by extreme conductor bars forming thus a rail.

Moreover, the magnetic field (similar to a magnet, for instance) moves rapidly above this group (Fig. 2.1).

According to Faraday's law, a voltage is induced across each conductor cut by the magnetic field (expression 1.23):

$$e = -B\,l\,v \tag{2.1}$$

A first consequence appears. As each conductor is short-circuited, an eddy current i starts to circulate into the conductor, which is momentarily under the magnetic field (or the magnet). As this current crosses the magnetic field, according to Laplace's law, a magnetic force (Eq. 1.5) is carried out on this conductor. This force leads the conductor who follows the magnetic field's movement. If those conductors are mobile they speed up and while they are gaining speed, the speed in which the magnetic field is cut by those conductors slows down and the induced voltage decreases, as well as current i. This effect of Lenz's law results in decreasing Laplace's force. Thus if the conductors moved at the same speed as the magnetic field, the induced voltages, the current i and the force would be canceled out. The rotor speed is then slightly inferior to the magnetic field's speed.

In a cage induction machine, the rail considered in the example of the Fig. 2.1 is curved to form the squirrel cage and the magnetic field's movement becomes a rotating field, created by three stator windings. The stator windings are supplied by a three-phase system of balanced voltages with the pulsation ω_s, crossed by a three-phase current system with the same pulsation and generate stator fluxes. We define the pulsation as the electrical speed $\omega_s = 2\pi f_s$ with f_s the frequency of the stator voltage in Hz. According to Ferraris's theorem (Caron and Hautier 1995), a rotating magnetic field is created in the air gap (and curls itself in the rotor and stator framework). Its angular speed, which is called synchronous speed, equals the pulsation of the three-phase system balanced with currents that run through these windings. For the stator of the studied machine, the magnetic field generated by the stator rotates by one spin during an electrical quantity period. In the case where the stator is composed of a number p of pole pairs by phase, the stator magnetic field is rotating at the synchronous speed (in rad/s): $\Omega_s = \omega_s/p$.

As previously explained, the speed of the rotor is inferior to the synchronous speed in permanent operating: $\Omega \neq \Omega_s$. The rotor conductors are then submitted to a variable magnetic field that rotates in relation to themselves at relative speed $\Omega_r = \Omega_s - \Omega$.

This relative speed induces the voltage (Faraday's law) in each closed loop of the rotor conductors with pulsation: $\omega_r = p\Omega_r$.

The stator windings are in short circuits for cage machines or are connected to an external supply circuit for Doubly-Fed Induction Machine. Hence, the induced e.m.f. will generate rotor currents of the same pulsation in the closed rotor circuit. These currents then create a rotor magnetic field that rotates, in relation to the rotor, at speed $\Omega_r = \omega_r/p$.

Given that the rotor turns at speed Ω, the rotor field speed in relation to the stator is $\Omega_r + \Omega = \Omega_s$.

The magnetic field generated by the rotor windings and the magnetic field generated by the stator windings then rotate at the same synchronous speed and are combined to create a "resulting magnetic field" in the air gap. Therefore, the physical phenomena (in physics) generated by the stator circuit, including those induced by the rotor circuit will generate electrical quantities at the stator circuit at ω_s pulsation level. The electrical quantities specific to rotor circuit will all be at ω_r pulsation.

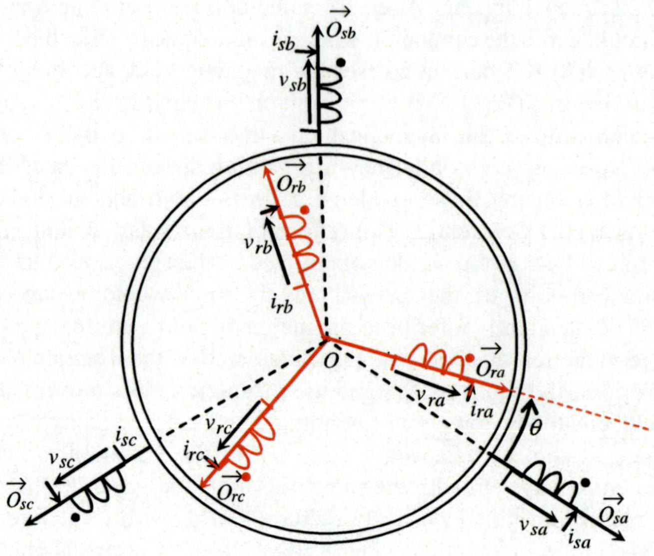

Fig. 2.2 Three-phase windings in the "natural frame"

2.1.3 Equivalent Representation and Vector Formulation

The studied machine is described in the electrical space by:

- Three identical windings for the stator phases set on the stator length and whose axis are distant two by two of an electrical angle equal to $2\pi/3$.
- Three rotor windings whose axis are equally distant between themselves of an electrical angle equal to $2\pi/3$, rotating at the mechanic speed Ω. In the case of a short-circuited rotor machine, those windings strictly account for the cage behavior.

The Fig. 2.2 represents the axis position of the stator ($\vec{O}_{\text{sa}}$, $\vec{O}_{\text{sb}}$, $\vec{O}_{\text{sc}}$) and rotor ($\vec{O}_{\text{ra}}$, $\vec{O}_{\text{rb}}$, $\vec{O}_{\text{rc}}$) phases in the electrical space (the electrical angle θ equals to the mechanical angle multiplied by the number p of pole pairs per phase). The stator quantities are respectively written under vector form:

$$[V_s] = \begin{bmatrix} v_{\text{sa}} \\ v_{\text{sb}} \\ v_{\text{sc}} \end{bmatrix}, [I_s] = \begin{bmatrix} i_{\text{sa}} \\ i_{\text{sb}} \\ i_{\text{sc}} \end{bmatrix}, [\phi_s] = \begin{bmatrix} \phi_{\text{sa}} \\ \phi_{\text{sb}} \\ \phi_{\text{sc}} \end{bmatrix} \tag{2.2}$$

For the rotor windings, the following voltage, current and flux vectors will be taken into account:

$$[V_r] = \begin{bmatrix} v_{\text{ra}} \\ v_{\text{rb}} \\ v_{\text{rc}} \end{bmatrix}, [I_r] = \begin{bmatrix} i_{\text{ra}} \\ i_{\text{rb}} \\ i_{\text{rc}} \end{bmatrix}, [\phi_r] = \begin{bmatrix} \phi_{\text{ra}} \\ \phi_{\text{rb}} \\ \phi_{\text{rc}} \end{bmatrix} \tag{2.3}$$

If the considered machine is a cage machine, the rotor voltages' vector is null. The only power supply of the stator windings creates simultaneously the rotating magnetic field and the induced currents into the rotor windings.

2.2 Dynamic Modeling in a Three-Axe Frame

2.2.1 Hypotheses

The modeling of the three-phase induction machine generally retained, relies on several hypotheses which are now recalled [Caron and Hautier 1995].

The first hypothesis consists in assuming that the magnetomotive forces created by different stator and rotor phases are spread in a sinusoidal way along the air gap, when those windings are crossed by a constant current. An appropriate dispersion of the windings in space allows reaching this aim.

The machine air gap is also supposed to be constantly thick: the notching effects, generating space harmonic, are ignored. These hypotheses will allow restricting the modeling of fundamental components (low frequency) of alternative quantities.

For this modeling, hypotheses about the physical behavior of the materials are expressed:

- The magnetic fields are not saturated, are not submitted to the hysteresis phenomenon and are not the center of Foucault's currents (for all practical purposes, the magnetic circuit is leafed through to limit the effects). This allows defining linear inductions.
- The skin effect is not taken into account.
- The temperature in the motor stays constant whatever the operating point is, which leads to constant parameters in mathematical models (stationarity).

These hypotheses will allow adding the associated fluxes to the different currents, using proper constant inductions, characterizing couplings by sinusoidal variations of the mutual inductions and representing induction flows by a spatial vector. They allow the development of modeling with a limited complexity, and thus the development of control strategies which can be implemented in practice.

In fact, the parameters of these models vary in saturation, skin effect and temperature. These variations' influences on the different control strategies' performances are analyzed in Chap. 4.

2.2.2 Study of the Electromechanical Conversion

2.2.2.1 Electrical Conversions

Finally, the hypotheses allow the electrical and magnetic behaviors to be considered as totally linear. The total variation of the energy exchanged with the electrical supply can be divided in two terms linked with stator and rotor circuits (paragraph 1.4.1):

$$\mathrm{d}W_{\mathrm{et}} = \mathrm{d}W_{\mathrm{et\_s}} + \mathrm{d}W_{\mathrm{et\_r}} \tag{2.4}$$

Those terms are expressed in matrix form as follows:

$$\mathrm{d}W_{\mathrm{et\_s}} = [I_s]^T\,[V_s]\;\mathrm{d}t \tag{2.5}$$

$$\mathrm{d}W_{\mathrm{et\_r}} = [I_r]^T\,[V_r]\;\mathrm{d}t \tag{2.6}$$

It must be retained that, in order to facilitate the understanding, all the matrix and vector quantities will be written between brackets. The total variation of energy dispersed in Joule losses is written:

$$dW_{\mathrm{pertesJoule}} = [I_s]^T[R_s][I_s]\mathrm{d}t + [I_r]^T[R_r][I_r]\mathrm{d}t \tag{2.7}$$

with

$$[R_s] = \begin{bmatrix} R_s & 0 & 0 \\ 0 & R_s & 0 \\ 0 & 0 & R_s \end{bmatrix} \text{ and } [R_r] = \begin{bmatrix} R_r & 0 & 0 \\ 0 & R_r & 0 \\ 0 & 0 & R_r \end{bmatrix} \tag{2.8}$$

where:

R_s is a stator winding resistor,
R_r is a rotor winding resistor.

The iron losses being ignored, the electromechanical energy variation stored up in the machine is then:

$$\mathrm{d}W_e = \mathrm{d}W_{\mathrm{et}} - \mathrm{d}W_{\mathrm{Joulelosses}} \tag{2.9}$$

The variation of the electromagnetic energy is then written:

$$\mathrm{d}W_e = [I_s]^T\mathrm{d}[\phi_s] + [I_r]^T\mathrm{d}[\phi_r] \tag{2.10}$$

By identifying the Eq. 2.9 and integrating (2.7), (2.4), (2.5) and (2.6), Faraday's law is to be found again, which expresses the relation between voltages across the windings resistors' terminals, the currents and the total flow variations in their respective frames:

$$\left\{\frac{\mathrm{d}[\phi_s]}{\mathrm{d}t} = [V_s] - [R_s][I_s]\right\}_{\mathrm{stator}} \tag{2.11}$$

$$\left\{\frac{\mathrm{d}[\phi_r]}{\mathrm{d}t} = [V_r] - [R_r][I_r]\right\}_{\mathrm{rotor}} \tag{2.12}$$

where:

$[\phi_s]$ is the vector gathering the magnetic flows captured by the stator windings,
$[\phi_r]$ is the vector gathering the magnetic flows captured by the rotor windings.

Both equations are a matrix generalization of the Eq. 1.59. After integrating each differential equation, the expression of the flux is obtained in its own frame (stator or rotor).

The induction of each winding is divided into a main induction (or proper, written p) that participates in the common flux and a leak induction (written σ). We then define:

- the proper induction of a stator winding:

$$l_s = l_{\mathrm{sp}} + l_{s\sigma} \tag{2.13}$$

- the proper induction of a rotor winding:

$$l_r = l_{\mathrm{rp}} + l_{r\sigma} \tag{2.14}$$

Without saturation, the fluxes are supposed to be dependent of the currents in a linear way. It must be kept in mind that there are six windings magnetically coupled, three of which are mobile. The total flux in each winding is given by the sum of its proper flux (linked by the induction l_s, for a stator flux) with three rotor-coupling fluxes (linked by mutual inductions variable according to the rotor position). For a stator flux, we then obtain:

$$\begin{aligned}\phi_{\mathrm{sa}} &= l_s i_{\mathrm{sa}} + M_s i_{\mathrm{sb}} + M_s i_{\mathrm{sc}} \\ &\quad + M_{\mathrm{sr}}\left(i_{\mathrm{ra}}\cos(\theta) + i_{\mathrm{rb}}\cos\left(\theta - \frac{2\pi}{3}\right) + i_{\mathrm{rc}}\cos\left(\theta - \frac{4\pi}{3}\right)\right)\end{aligned} \tag{2.15}$$

The mutual induction between two distinct stator phases is written as follows:

$$M_s = l_{\mathrm{sp}}\cos\left(\frac{2\pi}{3}\right) = -\frac{1}{2}l_{\mathrm{sp}} \tag{2.16}$$

M_{sr} is the maximal mutual induction between a stator phase and a rotor phase when their axes are collinear. Given the spatial structure of the machine, the mutual inductions between the stator phase and the rotor phase are written as follows:

$$\begin{aligned}M_{\mathrm{sr}}\cos\left(\vec{O}_{\mathrm{sa}}, \vec{O}_{\mathrm{ra}}\right) &= M_{\mathrm{sr}}\cos(\theta) \\ M_{\mathrm{sr}}\cos\left(\vec{O}_{\mathrm{sa}}, \vec{O}_{\mathrm{rb}}\right) &= M_{\mathrm{sr}}\cos\left(\theta - \frac{2\pi}{3}\right) = M_{\mathrm{sr}}\cos\left(\theta + \frac{4\pi}{3}\right) \\ M_{\mathrm{sr}}\cos\left(\vec{O}_{\mathrm{sa}}, \vec{O}_{\mathrm{rc}}\right) &= M_{\mathrm{sr}}\cos\left(\theta - \frac{4\pi}{3}\right) = M_{\mathrm{sr}}\cos\left(\theta + \frac{2\pi}{3}\right)\end{aligned}$$

where θ is the electrical angle between a rotor phase and a stator phase of the same name and defined as such in Fig. 2.2.

The other stator fluxes' expressions are obtained by circularly permuting the Eq. 2.15. Thus, under vector form, a generalization at the three-phase case of the Eq. 1.51 of the elementary machine described in Chap. 1 is obtained:

$$\left\{ [\phi_s] = \begin{bmatrix} \phi_{sa} \\ \phi_{sb} \\ \phi_{sc} \end{bmatrix} = \begin{bmatrix} l_s & M_s & M_s \\ M_s & l_s & M_s \\ M_s & M_s & l_s \end{bmatrix} \begin{bmatrix} i_{sa} \\ i_{sb} \\ i_{sc} \end{bmatrix} + M_{sr}[R(\theta)] \begin{bmatrix} i_{ra} \\ i_{rb} \\ i_{rc} \end{bmatrix} \right\}_{\text{stator}} \tag{2.17}$$

$[R(\theta)]$ is a rotational matrix that allows projecting rotor quantities in the stator frame according to the rotor position in relation to the stator:

$$[R(\theta)] = \begin{bmatrix} \cos(\theta) & \cos\left(\theta - \frac{4}{3}\pi\right) & \cos\left(\theta - \frac{2}{3}\pi\right) \\ \cos\left(\theta - \frac{2}{3}\pi\right) & \cos(\theta) & \cos\left(\theta - \frac{4}{3}\pi\right) \\ \cos\left(\theta - \frac{4}{3}\pi\right) & \cos\left(\theta - \frac{2}{3}\pi\right) & \cos(\theta) \end{bmatrix} \tag{2.18}$$

To determine the equations of the rotor circuit, the mutual induction between two distinct rotor phases is defined:

$$M_r = l_{rp} \cos\left(\frac{2\pi}{3}\right) = -\frac{1}{2} l_{rp} \tag{2.19}$$

The rotor fluxes are obtained by keeping the same reasoning:

$$\left\{ [\phi_r] = \begin{bmatrix} \phi_{ra} \\ \phi_{rb} \\ \phi_{rc} \end{bmatrix} = \begin{bmatrix} l_r & M_r & M_r \\ M_r & l_r & M_r \\ M_r & M_r & l_r \end{bmatrix} \begin{bmatrix} i_{ra} \\ i_{rb} \\ i_{rc} \end{bmatrix} + M_{sr}[R(\theta)]^{\mathrm{T}} \begin{bmatrix} i_{sa} \\ i_{sb} \\ i_{sc} \end{bmatrix} \right\}_{\text{rotor}} \tag{2.20}$$

Each relation is given again in its respective frame. The windings being coupled in star, the three-phase current systems are balanced, and it results in:

$$i_{sa} + i_{sb} + i_{sc} = 0 \tag{2.21}$$

$$i_{ra} + i_{rb} + i_{rc} = 0 \tag{2.22}$$

By expressing the winding current c depending on the two others, each phase's flux is simplified. For instance, for the stator flux of phase a, the flux is given by its own flux linked by $(l_s + M_s)$ and the coupling fluxes:

$$\phi_{sa} = (l_s - M_s) i_{sa} + M_{sr}\left(i_{ra}\cos(\theta) + i_{rb}\cos\left(\theta - \frac{2\pi}{3}\right) + i_{rc}\cos\left(\theta - \frac{4\pi}{3}\right)\right) \tag{2.23}$$

By taking the other fluxes into account, the following matrix form is obtained:

$$\left\{ [\phi_s] = \begin{bmatrix} \phi_{sa} \\ \phi_{sb} \\ \phi_{sc} \end{bmatrix} = \begin{bmatrix} L_s & 0 & 0 \\ 0 & L_s & 0 \\ 0 & 0 & L_s \end{bmatrix} \begin{bmatrix} i_{sa} \\ i_{sb} \\ i_{sc} \end{bmatrix} + M_{sr}[R(\theta)] \begin{bmatrix} i_{ra} \\ i_{rb} \\ i_{rc} \end{bmatrix} \right\}_{\text{stator}} \tag{2.24}$$

$$\left\{ [\phi_r] = \begin{bmatrix} \phi_{ra} \\ \phi_{rb} \\ \phi_{rc} \end{bmatrix} = \begin{bmatrix} L_r & 0 & 0 \\ 0 & L_r & 0 \\ 0 & 0 & L_r \end{bmatrix} \begin{bmatrix} i_{ra} \\ i_{rb} \\ i_{rc} \end{bmatrix} + M_{sr}[R(\theta)]^T \begin{bmatrix} i_{sa} \\ i_{sb} \\ i_{sc} \end{bmatrix} \right\}_{\text{rotor}} \tag{2.25}$$

The cyclic stator inductance per phase, taking into account the contribution of the three stator windings (at the creation of a stator flux), is written as:

$$L_s = l_s - M_s = \frac{3}{2} l_{sp} + l_{s\sigma} \tag{2.26}$$

The inner cyclic inductance, l_{sp}, is still called magnetizing inductance since it creates the flux into the air gap without current in the rotor circuit.

The rotor cyclic inductance per phase, taking into account the contribution of the three rotor windings (at the creation of a rotor flux), is written as follows:

$$L_r = l_r - M_r = \frac{3}{2} l_{rp} + l_{r\sigma} \tag{2.27}$$

The relations between flux and currents can be condensed using particular matrices:

$$\begin{bmatrix} [\phi_s] \\ [\phi_r] \end{bmatrix} = \begin{bmatrix} [L_s] & [M_{sr}(\theta)] \\ [M_{sr}(\theta)]^{\mathrm{T}} & [L_r] \end{bmatrix} \begin{bmatrix} [I_s] \\ [I_r] \end{bmatrix} \tag{2.28}$$

with the matrices defined as follows:

$$[L_s] = \begin{bmatrix} L_s & 0 & 0 \\ 0 & L_s & 0 \\ 0 & 0 & L_s \end{bmatrix}, \ [L_r] = \begin{bmatrix} L_r & 0 & 0 \\ 0 & L_r & 0 \\ 0 & 0 & L_r \end{bmatrix}, \ [M_{sr}(\theta)] = M_{sr}[R(\theta)]$$

2.2.2.2 Electromagnetic Transformation

By replacing flux expressions (2.24) and (2.25) in the relation of electromagnetic energy variation (2.10), the following formula is obtained:

$$\begin{aligned} \mathrm{d}W_e = {} & [I_s]^T [L_s]\mathrm{d}[I_s] + [I_s]^T M_{sr}\mathrm{d}([R(\theta)] [I_r]) \\ & + [I_r]^T [L_r]\mathrm{d}[I_r] + [I_r]^T M_{sr}\mathrm{d}\big([R(\theta)]^T [I_s]\big) \end{aligned} \tag{2.29}$$

By developing the product's derivatives, the following formula is obtained:

$$\begin{aligned} \mathrm{d}W_e = {} & [I_s]^T [L_s]\mathrm{d}[I_s] + [I_s]^T M_{sr}\mathrm{d}[R(\theta)] [I_r] + M_{sr}[I_s]^T [R(\theta)] \mathrm{d}[I_r] \\ & + [I_r]^T [L_r]\mathrm{d}[I_r] + [I_r]^T M_{sr}\mathrm{d}[R(\theta)] [I_s] + M_{sr}[I_r]^T [R(\theta)] \mathrm{d} [I_s] \end{aligned} \tag{2.30}$$

$$\begin{aligned}\mathrm{d}W_e = {} & [I_s]^T\,[L_s]\,\mathrm{d}[I_s] + [I_r]^T\,[L_r]\,\mathrm{d}[I_r] + [I_s]^T 2M_{\mathrm{sr}}\mathrm{d}[R(\theta)]\,[I_r] \\ & + [I_s]^T M_{\mathrm{sr}}[R(\theta)]\,\mathrm{d}[I_r] + [I_r]^T\,M_{\mathrm{sr}}[R(\theta)]\,\mathrm{d}\,[I_s]\end{aligned} \tag{2.31}$$

along with

$$\mathrm{d}[R(\theta)] = -\begin{bmatrix} \sin(\theta) & \sin\left(\theta - \frac{4}{3}\pi\right) & \sin\left(\theta - \frac{2}{3}\pi\right) \\ \sin\left(\theta - \frac{2}{3}\pi\right) & \sin(\theta) & \sin\left(\theta - \frac{4}{3}\pi\right) \\ \sin\left(\theta - \frac{4}{3}\pi\right) & \sin\left(\theta - \frac{2}{3}\pi\right) & \sin(\theta) \end{bmatrix}\mathrm{d}\theta \tag{2.32}$$

Further information on matrix computation can be found in [Rotella and Borne 1995].

When the angular variation is null ($\theta = 0$), the accumulated magnetic energy is immediately induced as follows:

$$W_{\mathrm{mag}} = \frac{1}{2}\,[I_s]^T\,[L_s]\,[I_s] + \frac{1}{2} \times\,[I_r]^T\,[L_r]\,[I_r] + [I_s]^T M_{\mathrm{sr}}\mathrm{d}[R(\theta)]^T\,[I_r] \tag{2.33}$$

A generalized expression of Eq. 1.54 is found again. Therefore, the total variation of the magnetic energy is expressed as follows:

$$\begin{aligned}\mathrm{d}W_{\mathrm{mag}} = {} & [I_s]^T\,[L_s]\,\mathrm{d}[I_s] + [I_r]^T\,[L_r]\,\mathrm{d}[I_r] \\ & + \mathrm{d}[I_s]^T M_{\mathrm{sr}}[R(\theta)]\,[I_r] + [I_s]^T M_{\mathrm{sr}}\mathrm{d}[R(\theta)] \times [I_r] + [I_s]^T M_{\mathrm{sr}}[R(\theta)]\,\mathrm{d}[I_r]\end{aligned} \tag{2.34}$$

When ignoring the iron losses, the generalization of relation (1.47) results in:

$$\mathrm{d}W_m = \mathrm{d}W_e - \mathrm{d}W_{\mathrm{mag}} \tag{2.35}$$

That is, by calculating the difference between (2.31) and (2.34) for a pair of poles:

$$\mathrm{d}W_m = [I_s]^T M_{\mathrm{sr}}\mathrm{d}[R(\theta)]\,[I_r] \tag{2.36}$$

The electrical energy variation corresponds to the mechanical energy variation:

$$\mathrm{d}W_m = T\mathrm{d}\theta \tag{2.37}$$

Therefore, for a number p of pole pairs, the torque is expressed by integrating (2.36) into (2.37):

$$T = p[I_s]^T M_{\mathrm{sr}}\frac{\mathrm{d}[R(\theta)]}{\mathrm{d}\theta}\,[I_r] \tag{2.38}$$

or the following transposed expression:

$$T = p[I_r]^T M_{\mathrm{sr}}\frac{\mathrm{d}[R(\theta)]}{\mathrm{d}\theta}\,[I_s] \tag{2.39}$$

Table 2.1 Causal organization of the mathematical equations for the induction machine

Electrical relation:	
$R1_s$: $\{[V_{Rs}] = [R_s][I_s]\}_{\text{stator}}$	$R1_r$:$\{[V_{Rr}] = [R_r][I_r]\}_{\text{rotor}}$
$R2_s$: $\{[e_s] = [V_s] - [V_{Rs}]\}_{\text{stator}}$	$R2_r$: $\{[e_r] = [V_r] - [V_{Rr}]\}_{\text{rotor}}$
$R3_s$: $\left\{[\phi_s] = \int_{t0}^{t0+\Delta t} [e_s]\,\text{dt} + [e_s](t_0)\right\}_{\text{stator}}$	$R3_r$: $\left\{[\phi_r] = \int_{t0}^{t0+\Delta t} [e_r]\,\text{dt} + [e_r](t_0)\right\}_{\text{rotor}}$
Flow/current coupling: *R4:* (2.28) inverse relation	
Electromechanical conversion: $R5$: $T = p[I_s]^T M_{\text{sr}} \frac{\text{d}[R(\theta)]}{\text{d}\theta}[I_r]$	

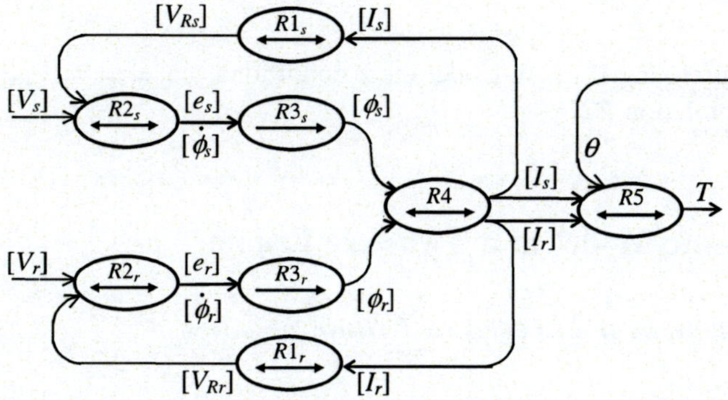

Fig. 2.3 COG of the three-phase model for the doubly-fed induction machine

or any linear combination of these two expressions, for instance [Grenier et al. 2001]:

$$T = \frac{1}{2} p M_{\text{sr}} \left([I_s]^{\text{T}} \frac{\text{d}[R(\theta)]}{\text{d}\theta} [I_r] + [I_r]^{\text{T}} \frac{\text{d}[R(\theta)]}{\text{d}\theta} [I_s] \right) \tag{2.40}$$

2.2.2.3 Causal Ordering Graph of the Model

Stator fluxes are obtained through the integration of matrix Eq. 2.11. In this equation, Joule effect losses are modeled by a voltage drop at the terminals of a resistor, for which the relation is $R1_s$ (Table 2.1). The mesh law, $R2_s$, allows determining the e.m.f. depending on the voltage drops and on the voltages applied on the stator terminals. Integrating this equation, $R3_s$, allows calculating the stator fluxes. The same can be done for the rotor equations. Currents can subsequently be determined by the inverse equations in the coupling with the fluxes (Eq. 2.28) *R*4.

Expressions for the torque (2.38), for the rotation matrix and its derivative are directly useable: relation *R*5. Figure 2.3 represents the causal ordering graph for the model of the coiled rotor induction machine. The mechanical part, which highly depends on the load, is not represented. The induction cage machine model

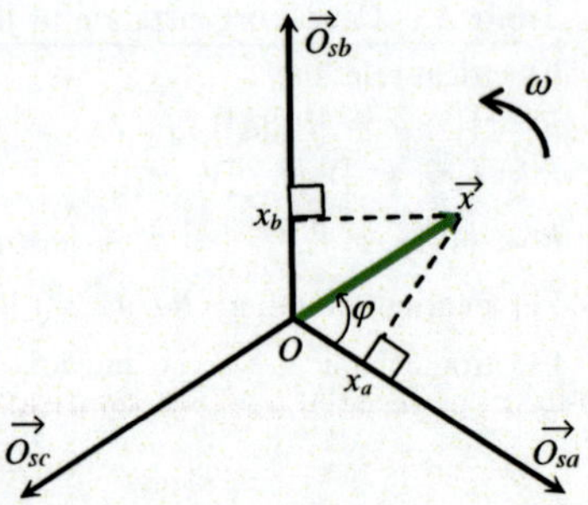

Fig. 2.4 Representation of a rotating phasor in a fixed three-axe frame

is obtained by fixing $[V_r] = 0$. Under these conditions, since $[e_r] = [V_{\mathrm{Rr}}]$, there is no need to use relation $R2_{\mathrm{r}}$.

2.3 Dynamic Model in a Two-Axe Frame

2.3.1 Phasor in a Three-Axe Frame

One way to make the established mathematical model less complex is to describe the induction machine by taking into account two (equivalent) windings rather than three. This method relies on working out the equation of a phasor and will be explained in the following paragraphs.

A three-phase winding supplied by a three-phase currents system creates three-phase pulsating magnetic fields. According to the Ferraris theorem [Caron and Hautier 1995], a rotating magnetic field appears in the air gap of the machine and results from the spatial combination of the fields. Voltages, currents and stator, as well as rotor three-phase flows create as many phasors in the three-phase space defined in Fig. 2.2. Therefore, a three-phase system is represented with the following vector form

$$[X] = \begin{bmatrix} x_a \\ x_b \\ x_c \end{bmatrix}.$$

It includes three variables x_a, x_b, x_c, which are made to correspond to projections of a spatial vector $\vec{x}$ on the three axes directed by unit vectors $\vec{O}_{\mathrm{sa}}, \vec{O}_{\mathrm{sb}}, \vec{O}_{\mathrm{sc}}$ out of phase with $2/3\,\pi$ (Fig. 2.4).

$$\vec{x} = K\left(x_a\,\vec{O}_{\mathrm{sa}} + x_b\,\vec{O}_{\mathrm{sb}} + x_c\,\vec{O}_{\mathrm{sc}}\right) \tag{2.41}$$

where K is a constant.

The phasor corresponds to a Fresnel vector rotating at angular speed ω and can thus be represented with a complex number:

$$\underline{x} = K\begin{bmatrix} 1 & e^{j\frac{2}{3}\pi} & e^{j\frac{4}{3}\pi} \end{bmatrix}\begin{bmatrix} x_a \\ x_b \\ x_c \end{bmatrix} \tag{2.42}$$

To enhance readability, each complex value is underlined.

We can for instance consider a three-phase system with a balanced current, a pulsation equal to ω and a root-mean-square-value equal to I:

$$[I] = \begin{bmatrix} i_a \\ i_b \\ i_c \end{bmatrix} = I\sqrt{2}\begin{bmatrix} \sin(\omega t) \\ \sin\left(\omega t - \frac{2\pi}{3}\right) \\ \sin\left(\omega t - \frac{4\pi}{3}\right) \end{bmatrix} \tag{2.43}$$

The associated complex number is:

$$\underline{i} = I\sqrt{3}\mathrm{e}^{j\omega t} \tag{2.44}$$

The inverse relations are subsequently expressed according to:

$$[X] = \begin{bmatrix} x_a \\ x_b \\ x_c \end{bmatrix} = \frac{2}{3K}\begin{bmatrix} \mathrm{Real}(x) \\ \mathrm{Real}\left(xe^{-j\frac{2\pi}{3}}\right) \\ \mathrm{Real}\left(xe^{-j\frac{4\pi}{3}}\right) \end{bmatrix} \tag{2.45}$$

where Real(...) represents the real part of the expression between brackets.

When $K = 2/3$, a vector representation that preserves the amplitudes is obtained. When $K = \sqrt{2/3}$, the obtained representation preserves the power. This value is the chosen one for the modeling presentation in this chapter.

2.3.2 Phasor in a Fixed Two-Axe Frame

In an orthogonal frame $\left(\vec{O}_\alpha, \vec{O}_\beta\right)$, aligned on axe $\left(\vec{O}_{sa}\right)$, this same vector will be expressed by means of (Fig. 2.5):

$$\vec{x} = x_\alpha \vec{O}_{s\alpha} + x_\beta \vec{O}_{s\beta} \tag{2.46}$$

Those coordinates may be obtained directly from the coordinates in the three-axe frame by using a rotation matrix:

$$\begin{bmatrix} x_\alpha \\ x_\beta \end{bmatrix} = \sqrt{\frac{2}{3}}\begin{bmatrix} \cos(0) & \cos\left(\frac{2\pi}{3}\right) & \cos\left(\frac{4\pi}{3}\right) \\ \sin(0) & \sin\left(\frac{2\pi}{3}\right) & \sin\left(\frac{4\pi}{3}\right) \end{bmatrix}\begin{bmatrix} x_a \\ x_b \\ x_c \end{bmatrix}$$

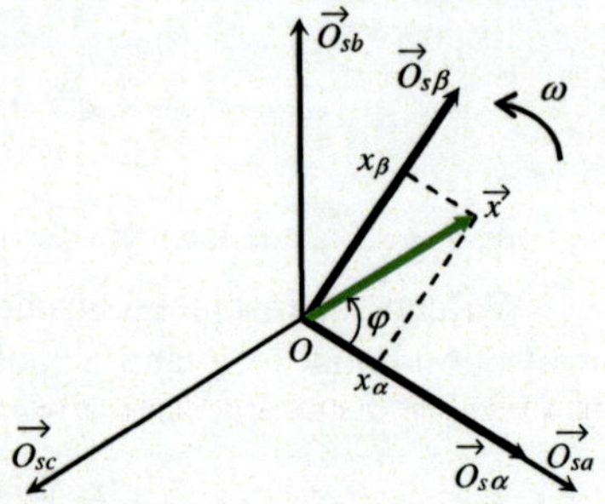

Fig. 2.5 Representation of a phasor in a two-axe frame

$$\begin{bmatrix} x_\alpha \\ x_\beta \end{bmatrix} = \sqrt{\frac{2}{3}} \begin{bmatrix} 1 & -\frac{1}{2} & -\frac{1}{2} \\ 0 & \frac{\sqrt{3}}{2} & -\frac{\sqrt{3}}{2} \end{bmatrix} \begin{bmatrix} x_a \\ x_b \\ x_c \end{bmatrix} \tag{2.47}$$

If this same principle is applied to the magnetic field created by three-phase windings (Chap. 2, paragraph 2.1.3), this same field can also be obtained with two coils, the axes of which are perpendicular and are supplied by a two-phase electric system.

If an angle θs exists between the frame and the synchronous frame, it is necessary to take this angle into account:

$$\vec{x} = \underline{x}\, e^{j(\theta s)}\, \vec{O}_{s\alpha} \tag{2.48}$$

2.3.3 Transformation Matrices

An equivalent rotating magnetic field can also be created by taking into account only two phases. The Concordia transformation makes it possible to obtain an equivalent system formed by three orthogonal windings (axes). Two of these windings are located in the same plan as the three-phase windings. The third winding is located in the orthogonal plan formed by the three-phase windings. It represents the homopolar component, which characterizes the balance of the studied system:

$$x_0 = \frac{1}{\sqrt{3}}\,(x_a + x_b + x_c) \tag{2.49}$$

This homopolar component is null when the system is balanced. The passage between coordinates in the three-axe frame, as well as two-axe and homopolar coordinates is defined in terms of:

$$\begin{bmatrix} x_\alpha \\ x_\beta \\ x_0 \end{bmatrix} = \sqrt{\frac{2}{3}} \begin{bmatrix} 1 & -\frac{1}{2} & -\frac{1}{2} \\ 0 & \frac{\sqrt{3}}{2} & -\frac{\sqrt{3}}{2} \\ \frac{1}{\sqrt{2}} & \frac{1}{\sqrt{2}} & \frac{1}{\sqrt{2}} \end{bmatrix} \begin{bmatrix} x_a \\ x_b \\ x_c \end{bmatrix} \tag{2.50}$$

which can also be written as follows:

$$\left[X_{\alpha,\beta,0}\right] = [C]\,[X] \tag{2.51}$$

along with:

$$[C] = \sqrt{\frac{2}{3}} \begin{bmatrix} 1 & \frac{-1}{2} & \frac{-1}{2} \\ 0 & \frac{\sqrt{3}}{2} & \frac{-\sqrt{3}}{2} \\ \frac{1}{\sqrt{2}} & \frac{1}{\sqrt{2}} & \frac{1}{\sqrt{2}} \end{bmatrix} \tag{2.52}$$

This transformation allows going from a three-axe base to a two-axe orthogonal and orthonormal base. In literature, this is generally called Concordia inverse transform and it is written as follows: $[T_3]^{-1}(=[C])$. For instance, a three-phase current-balanced system, with a pulsation equal to $\omega = 2\pi 50$ rad/s and with a root-mean-square-value equal to $I = 8$A:

$$[I] = \begin{bmatrix} i_a \\ i_b \\ i_c \end{bmatrix} = I\sqrt{2} \begin{bmatrix} \sin(\omega t) \\ \sin\left(\omega t - \frac{2\pi}{3}\right) \\ \sin\left(\omega t - \frac{4\pi}{3}\right) \end{bmatrix} \tag{2.53}$$

The transformation leads to:

$$\left[I_{\alpha\beta 0}\right] = \begin{bmatrix} i_\alpha \\ i_\beta \\ i_0 \end{bmatrix} = [C] \begin{bmatrix} i_a \\ i_b \\ i_c \end{bmatrix} = \sqrt{\frac{2}{3}} I\sqrt{2} \begin{bmatrix} \sin(\omega t) \\ \sin\left(\omega t - \frac{\pi}{2}\right) \\ 0 \end{bmatrix} \tag{2.54}$$

The timing evolution of these values is represented on Fig. 2.6.

By means of the coordinates in the two-axe frame, the coordinates of a vector are found again in the three-axe frame by using the inverse matrix:

$$[X] = [C]^{-1}\left[X_{\alpha,\beta,0}\right] \tag{2.55}$$

along with:

$$[C]^{-1} = \sqrt{\frac{2}{3}} \begin{bmatrix} 1 & 0 & \frac{1}{\sqrt{2}} \\ -\frac{1}{2} & \frac{\sqrt{3}}{2} & \frac{1}{\sqrt{2}} \\ -\frac{1}{2} & -\frac{\sqrt{3}}{2} & \frac{1}{\sqrt{2}} \end{bmatrix} = [C]^{\mathrm{T}} \tag{2.56}$$

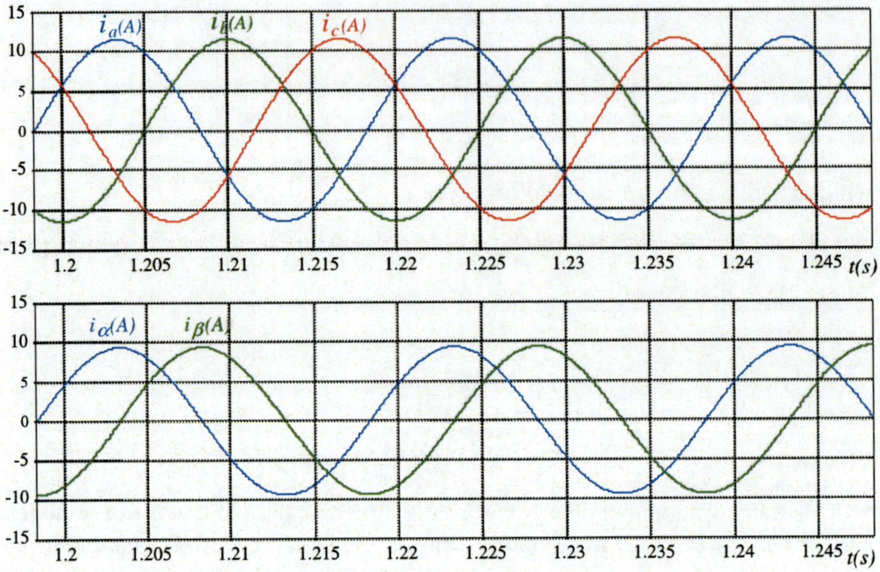

Fig. 2.6 Timing evolution of sinusoidal values in a three-axe frame of reference (*top* figure) and in a two-axe frame (*bottom* figure)

The Concordia inverse matrix is orthogonal and thus equals its transpose. In literature, this is generally called Condordia direct transform and it is written as follows: $[T_3]\left(=[C]^{-1}\right)$.

The instantaneous electrical power is expressed with:

$$p_{\text{et}}(t) = v_a i_a + v_b i_b + v_c i_c = [V]^T[I] \tag{2.57}$$

With the Concordia transposed, the power is expressed with:

$$p_{\text{et}}(t) = \left([C]^{-1}\left[V_{\alpha\beta 0}\right]\right)^T [C]^{-1}\left[I_{\alpha\beta 0}\right] = \left[V_{\alpha\beta 0}\right]^T[C]\,[C]^{-1}\left[I_{\alpha\beta 0}\right] = \left[V_{\alpha\beta 0}\right]^T\left[I_{\alpha\beta 0}\right] \tag{2.58}$$

The instantaneous electrical power is found again but is expressed with the Concordia components.

2.3.4 Vector Model in a Two-Axe Frame

2.3.4.1 Principle

At the stator, the three-phase voltages, the three-phase currents and the three-phase fluxes form three vectors rotating at the angular speed ω_s, depending on the fix stator: $\vec{V}_s$, $\vec{I}_s$, $\vec{\phi}_s$. At the rotor, the three-phase voltages, the three-phase currents and the three-phase fluxes, that rotate at the angular speed ω_r, depending on the rotor: $\vec{V}_r$, $\vec{I}_r$, $\vec{\phi}_r$. If the rotor spins at speed Ω, then these vectors rotate at the speed $p\Omega + \omega_r = \omega_s$ depending on the fix stator.

The use of the vector notations will allow the generalization of the obtained results with scalar values when modeling the elementary machine in paragraph 1.4.2.2.

By bringing the variables of three-axe frame (a,b,c) back on the axis of a two-axe frame (α, β), an equivalent two-axe machine can be taken into account. It is physically possible to create this two-phase machine.

2.3.4.2 Application to the Expressions of Fluxes

Working with the model of the induction machine on the three-axe frame, the aim is to determine the equivalent model on a two-axe frame. Flows in the three-axe frame (Eq. 2.28) are expressed according to the coordinates in the two-axe frame by applying Concordia inverse transformation (2.50) [Lesenne et al. 1995]:

$$[C]^{-1}\begin{bmatrix}[\phi_{s\_\alpha\beta 0}]\\ [\phi_{r\_\alpha\beta 0}]\end{bmatrix} = \begin{bmatrix}[L_s] & [M_{sr}(\theta)]\\ [M_{sr}(\theta)]^T & [L_r]\end{bmatrix}[C]^{-1}\begin{bmatrix}[I_{s\_\alpha\beta 0}]\\ [I_{r\_\alpha\beta 0}]\end{bmatrix} \tag{2.59}$$

Bringing all the Concordia transformations back into the right hand side, results in:

$$\begin{bmatrix}[\phi_{s\_\alpha\beta 0}]\\ [\phi_{r\_\alpha\beta 0}]\end{bmatrix} = \begin{bmatrix}[C]\,[L_s][C]^T & [C]\,[M_{sr}(\theta)][C]^T\\ [C]\,[M_{sr}(\theta)]^T[C]^T & [C]\,[L_r][C]^T\end{bmatrix}\begin{bmatrix}[I_{s\_\alpha\beta 0}]\\ [I_{r\_\alpha\beta 0}]\end{bmatrix} \tag{2.60}$$

The development of the previous formula leads to the expression of four submatrices, which define the coupling of the equivalent model in the two-axe frame.

$$\begin{bmatrix}[\phi_{s\_\alpha\beta 0}]\\ [\phi_{r\_\alpha\beta 0}]\end{bmatrix} = \begin{bmatrix}[L_{cs}] & [M_{csr}(\theta)]\\ [M_{csr}(\theta)]^T & [L_{cr}]\end{bmatrix}\begin{bmatrix}[I_{s\_\alpha\beta 0}]\\ [I_{r\_\alpha\beta 0}]\end{bmatrix} \tag{2.61}$$

along with:

$$[L_{cs}] = \begin{bmatrix}l_s - M_s & 0 & 0\\ 0 & l_s - M_s & 0\\ 0 & 0 & l_s + 2M_s\end{bmatrix} = \begin{bmatrix}L_s & 0 & 0\\ 0 & L_s & 0\\ 0 & 0 & L_{s0}\end{bmatrix} \tag{2.62}$$

$$[L_{cr}] = \begin{bmatrix}l_r - M_r & 0 & 0\\ 0 & l_r - M_r & 0\\ 0 & 0 & l_r + 2M_r\end{bmatrix} = \begin{bmatrix}L_r & 0 & 0\\ 0 & L_r & 0\\ 0 & 0 & L_{r0}\end{bmatrix} \tag{2.63}$$

$$[M_{csr}(\theta)] = \frac{3}{2}M_{sr}\,[R_c(\theta)] \tag{2.64}$$

along with:

$$[R_c(\theta)] = \begin{bmatrix} \cos(\theta) & -\sin(\theta) & 0 \\ \sin(\theta) & \cos(\theta) & 0 \\ 0 & 0 & 0 \end{bmatrix} \tag{2.65}$$

The formula uses several expressions with the following meanings:

- $L_s = l_s - M_s$ is the stator cyclic inductance,
- $L_r = l_r - M_r$ is the rotor cyclic inductance,
- $M = 3/2M_{sr}$ is the mutual cyclic inductance between a stator phase and a rotor phase when their axes are collinear (in the equivalent machine)
- $L_{so} = l_s + 2M_s$ and $L_{ro} = l_r + 2M_r$ correspond to homopolar inductances specific to each armature. If the magnetomotive forces have a sinusoidal repartition in space and if the rotor is smooth then the homopolar inductances L_{s0} and L_{r0} are null.

It can be noticed that the obtained model in the two-phase frame uses diagonal inductance matrices ($[L_{cs}]$ and $[L_{cr}]$). Therefore, the mutual inductances are null. This will bring important simplifications in the future calculations.

2.3.4.3 Application to Differential Equations

By including the Concordia transformation (Eq. 2.55) in the differential equations (Eqs. 2.1 and 2.12), results in:

$$\left\{ [C]^{-1}\frac{\mathrm{d}[\phi_{s\_\alpha\beta0}]}{\mathrm{d}t} = [C]^{-1}[V_{s\_\alpha\beta0}] - [R_s][C]^{-1}[I_{s\_\alpha\beta0}] \right\}_{\text{stator}} \tag{2.66}$$

$$\left\{ [C]^{-1}\frac{\mathrm{d}[\phi_{r\_\alpha\beta0}]}{\mathrm{d}t} = [C]^{-1}[V_{r\_\alpha\beta0}] - [R_r][C]^{-1}[I_{r\_\alpha\beta0}] \right\}_{\text{rotor}} \tag{2.67}$$

By multiplying on the left by using Concordia transformation, the expressions are simplified:

$$\left\{ \begin{array}{l} \frac{\mathrm{d}\phi_{s\alpha}}{\mathrm{d}t} = v_{s\alpha} - R_s i_{s\alpha} \\ \frac{\mathrm{d}\phi_{s\beta}}{\mathrm{d}t} = v_{s\beta} - R_s i_{s\beta} \\ \frac{\mathrm{d}\phi_{s0}}{\mathrm{d}t} = v_{s0} - R_s i_{s0} \end{array} \right\}_{\text{stator}} \tag{2.68}$$

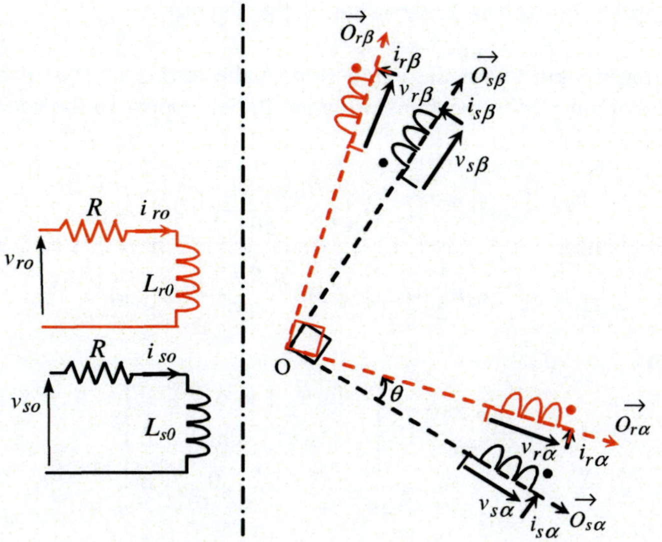

Fig. 2.7 Equivalent two-phase windings in an orthogonal frame

$$\left\{\begin{aligned}\frac{d\phi_{r\alpha}}{dt} &= v_{r\alpha} - R_r i_{r\alpha}\\ \frac{d\phi_{r\beta}}{dt} &= v_{r\beta} - R_r i_{r\beta}\\ \frac{d\phi_{r0}}{dt} &= v_{r0} - R_r i_{r0}\end{aligned}\right\}_{\text{rotor}} \tag{2.69}$$

When the sum of the three-phase components (*a, b, c*) is null, the third equation corresponding to the homopolar component is null (Eq. 2.49) and becomes void.

2.3.4.4 Equivalent Two-Phase Machine

The obtained electrical equations correspond to the equations of an equivalent two-phase induction machine. They are linked to two orthogonal two-axe frames as well as to the equations of two electrical circuits R-L connected to two homopolar machines (orthogonal to both two-axe frames). Figure 2.7 shows the position of the two-phase windings of the equivalent machine in both orthogonal frames (α, β). Axes $\vec{O}_{s\alpha}$ and $\vec{O}_{s\beta}$ are collinear because of the specific value of the Concordia matrix. Both homopolar circuits R-L are located orthogonally to these two two-phase windings.

2.3.4.5 Application to the Expression of the Torque

Starting from/working with the expression of the torque (2.38) in the three-axe frame, by making Concordia transformation (2.55) appear in the stator and rotor currents, results in:

$$T = p\left([C]^{-1}[I_{s\_\alpha\beta 0}]\right)^{\mathrm{T}} \frac{\mathrm{d}[M_{\mathrm{sr}}(\theta)]}{\mathrm{d}\theta}[C]^{-1}[I_{r\_\alpha\beta 0}] \tag{2.70}$$

$$T = p\frac{3}{2}[I_{s\_\alpha\beta 0}]^{\mathrm{T}} M_{\mathrm{sr}} \frac{\mathrm{d}[R_c(\theta)]}{\mathrm{d}\theta}[I_{r\_\alpha\beta 0}] \tag{2.71}$$

along with

$$\frac{d[R_c(\theta)]}{d\theta} = \begin{bmatrix} \sin(\theta) & \cos(\theta) & 0 \\ -\cos(\theta) & \sin(\theta) & 0 \\ 0 & 0 & 0 \end{bmatrix} \tag{2.72}$$

By writing:

$$M = \frac{3}{2}M_{\mathrm{sr}} \tag{2.73}$$

the scalar expression of the torque is obtained:

$$T = pM\left[\left(i_{\mathrm{r}\beta}i_{\mathrm{s}\alpha} - i_{\mathrm{r}\alpha}i_{\mathrm{s}\beta}\right)\cos(\theta) + \left(i_{\mathrm{r}\alpha}i_{\mathrm{s}\alpha} + i_{\mathrm{r}\beta}i_{\mathrm{s}\beta}\right)\sin(\theta)\right] \tag{2.74}$$

It is notable that only the mutual inductance contributes to the creation of the torque. Therefore, the homopolar components are only sources of losses.

2.3.4.6 Working With Balanced Currents

To cancel the homopolar components, the three-phase current system has to be balanced (Eqs. 2.21 and 2.22).

To impose a current-balanced system to the stator of a three-phase machine, it is possible to power the windings with only three phases, therefore without the connection of the neutral. The instantaneous sum of the stator currents is then null as well as the homopolar component of the current. For a balanced supply in terms of stator voltages, the homopolar component of the stator voltages is null too. If the homopolar component becomes void, all vectors will be reduced to two components.

$$\left\{ \frac{\mathrm{d}[\phi_{s\_\alpha\beta}]}{\mathrm{d}t} = [V_{s\_\alpha\beta}] - [R_s][I_{s\_\alpha\beta}] \right\}_{\text{stator}} \tag{2.75}$$

$$\left\{\frac{d[\phi_{r\_\alpha\beta}]}{dt} = [V_{r\_\alpha\beta}] - [R_r][I_{r\_\alpha\beta}]\right\}_{\text{rotor}} \tag{2.76}$$

with $[R_s] = R_s[I]$, $[R_r] = R_r[I]$ where $[I]$ is the identity matrix with a dimension of 2×2.

$$\begin{bmatrix} [\phi_{s\_\alpha\beta}] \\ [\phi_{r\_\alpha\beta}] \end{bmatrix} = \begin{bmatrix} [L_{cs}] & [M_{csr}(\theta)] \\ [M_{csr}(\theta)]^T & [L_{cr}] \end{bmatrix} \begin{bmatrix} [I_{s\_\alpha\beta}] \\ [I_{r\_\alpha\beta}] \end{bmatrix} \tag{2.77}$$

along with $[L_{cs}] = L_s[I]$, $[L_{cr}] = L_r[I]$ and

$$[R_c(\theta)] = \begin{bmatrix} \cos(\theta) & -\sin(\theta) \\ \sin(\theta) & \cos(\theta) \end{bmatrix}.$$

2.4 Vector Model in a Two-Axe Rotating Frame

2.4.1 The Park Transformation

The complex number associated to a balanced current system with root-mean-square value I and pulsation ω have been determined beforehand (Eq. 2.44). The same mathematical result can be obtained when supplying a coil with direct current and making it rotate at speed ω. This operation amounts to defining a rotating frame for axes d and q where the pulsation of every quantity is null.

$$\underline{i} = \underline{i}_{dq} e^{j\omega.t} \tag{2.78}$$

Under steady conditions, the electrical quantities shown in this frame are constant. The Concordia transformation and a rotation matrix will be used to establish this new frame. Rotation matrix $[R(\psi)]$ allows to bring the variables of (α, β, o) frame back to the axes of a (d,q,o) frame whose angle ψ may vary.

$$[R_P(\psi)] = \begin{bmatrix} \cos(\psi) & \sin(\psi) & 0 \\ -\sin(\psi) & \cos(\psi) & 0 \\ 0 & 0 & 1 \end{bmatrix} \tag{2.79}$$

If quantity ψ is time dependent, the frame will rotate. The product of the two frame changes (Concordia, rotation) defines the Park transformation of which basic property is to bring the stator and rotor quantities back to the same frame of reference. The matrix product that defines the Park transform is determined from:

$$[X_{dqo}] = [R_P(\psi)]\,[X_{\alpha\beta o}] \text{ et } [X_{\alpha\beta o}] = [C]\,[X_{abc}] \tag{2.80}$$

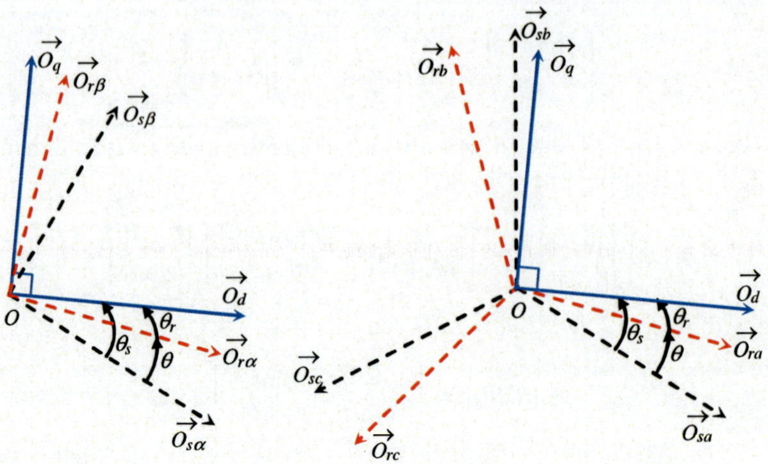

Fig. 2.8 Angular representation of the axis systems in the electrical space

The Park transformation is directly expressed by the following matrix product:

$$[X_{\text{dqo}}] = [P(\psi)]\,[X_{\text{abc}}] \tag{2.81}$$

along with:

$$[P(\psi)]=[R_P(\psi)]\,[C] \tag{2.82}$$

$$[P(\psi)] = \sqrt{\frac{2}{3}}\begin{bmatrix} \cos(\psi) & \cos\left(\psi - \frac{2\pi}{3}\right) & \cos\left(\psi - \frac{4\pi}{3}\right) \\ -\sin(\psi) & -\sin\left(\psi - \frac{2\pi}{3}\right) & -\sin\left(\psi - \frac{4\pi}{3}\right) \\ \frac{1}{\sqrt{2}} & \frac{1}{\sqrt{2}} & \frac{1}{\sqrt{2}} \end{bmatrix} \tag{2.83}$$

As matrices $[R_p(\psi)]$ and [C] are orthogonal, matrix $[P(\psi)]$ is orthogonal, too. Consequently, the inverse Park transformation is equal to its transposed and the power is retained [Caron and Hautier 1995]. It is relevant to note that the matrix defined as $[P(\psi)]$ is generally called "the inverse Park transformation". Consequently, its inverse is called "the direct Park transformation".

As an example, a current-balanced three-phase system with a pulsation $\omega = 2\pi 50$ rad/s and a root-mean-square-value $I = 8$A is studied again. When $\psi = \omega t$, applying the Park transformation leads to constant quantities depending on time of which the amplitude is proportional to the root-mean-square value:

$$\begin{bmatrix} i_d \\ i_q \\ i_0 \end{bmatrix} = \sqrt{3}I\begin{bmatrix} 1 \\ 0 \\ 0 \end{bmatrix} \tag{2.84}$$

2.4.2 Induction Machine Model in the Park Reference Frame

2.4.2.1 Principle

In paragraph 2.3.4, the induction machine was modeled using two separate frames. The first one is used to express stator quantities; the second one is used to express rotor quantities. Since these two frames are linked with angle θ, a model of the machine in a common frame named *d, q* can be obtained using the two rotation matrices.

Figure 2.8 shows the disposition of the two-phase or three-phase axis systems in the electrical space. At a certain point, the position of the magnetic field rotating in the air gap (paragraph 2.1.2) is pinpointed by angle θ_s, in relation to stationary axis $\vec{O}_{\text{sa}}$. For the development of the machine model, a Park reference frame is assumed to be lined up with this magnetic field and to rotate at the same speed (ω_s). Angle θ_s corresponds to the angle of axes $\vec{O}_{s\alpha}$ and θ_r; angle θ_r corresponds to the angle of axes $\vec{O}_{r\alpha}$ and $\vec{O}_d$. Transforming angle θ_s is necessary to bring the stator quantities back to the Park rotating reference frame. Transforming angle θ_r is necessary to bring the rotor quantities back. The figure indicates that the angles are linked by a relation in order to express the rotor and stator quantities in the same Park reference frame $(O,\, \vec{O}_d,\, \vec{O}_q)$. This relation is:

$$\theta_s = \theta + \theta_r \tag{2.85}$$

The same situation happens between the frame speeds in each frame and the mechanical speed, that is:

$$\omega_s = \omega + \omega_r, \tag{2.86}$$

with

$$\omega_s = \frac{d\theta_s}{dt},\quad \omega_r = \frac{d\theta_r}{dt},\quad \omega = p\Omega = \frac{d\theta}{dt} \tag{2.87}$$

Where Ω is the mechanical speed and ω this very speed viewed in the electrical space.

The speed of the rotor quantities is ω_r in relation to rotor speed ω. In relation to the stator frame, the rotor quantities consequently rotate at the same speed ω_s as the stator quantities. Using the Park transform will allow the conception of an induction machine model independent from the rotor position.

Two transformations are used. One $[P(\theta_s)]$ is applied to the stator quantities; the other $[P(\theta_r)]$ is applied to the rotor quantities.

$$\left[X_{\text{s\_dqo}}\right] = [P(\theta_s)]\,[X_{\text{s\_abc}}] \text{ et } \left[X_{\text{r\_dqo}}\right] = [P(\theta_r)]\,[X_{\text{r\_abc}}] \tag{2.88}$$

Direct and squared components x_d, x_q represent coordinates x_a, x_b, x_c in an orthogonal frame of reference rotating in the same plane. Term x_o represents the homopolar component, which is orthogonal to the plane constituted by the system x_a, x_b, x_c.

2.4.2.2 Determination of Differential Equations

Revealing the Park reference frame coordinates in the differential equations which are specific to the three-axe frame (Eqs. 2.11 and 2.12)—and applying the inverse Park transformation to the variables (flux, voltage and current)-, results in:

$$\frac{\mathrm{d}\left([P(\theta_s)]^{-1}[\phi_{\mathrm{s\_dq0}}]\right)}{\mathrm{dt}} = [P(\theta_s)]^{-1}[V_{\mathrm{s\_dq0}}] - [R_s][P(\theta_s)]^{-1}[I_{\mathrm{s\_dq0}}] \tag{2.89}$$

$$\frac{\mathrm{d}\left([P(\theta_r)]^{-1}[\phi_{\mathrm{r\_dq0}}]\right)}{\mathrm{dt}} = [P(\theta_r)]^{-1}[V_{\mathrm{r\_dq0}}] - [R_r][P(\theta_r)]^{-1}[I_{\mathrm{r\_dq0}}] \tag{2.90}$$

Multiplying on the left by the Park transformation, results in:

$$[P(\theta_s)]\frac{\mathrm{d}\left([P(\theta_s)]^{-1}\right)}{\mathrm{dt}}[\phi_{\mathrm{s\_dq0}}] + \frac{\mathrm{d}[\phi_{\mathrm{s\_dq0}}]}{\mathrm{dt}} = [V_{\mathrm{s\_dq0}}] - [R_s][I_{\mathrm{s\_dq0}}] \tag{2.91}$$

$$[P(\theta_r)]\frac{\mathrm{d}\left([P(\theta_r)]^{-1}\right)}{\mathrm{dt}}[\phi_{\mathrm{r\_dq0}}] + \frac{\mathrm{d}[\phi_{\mathrm{r\_dq0}}]}{\mathrm{dt}} = [V_{\mathrm{r\_dq0}}] - [R_r][I_{\mathrm{r\_dq0}}] \tag{2.92}$$

The product of the Park transformation with its derivative is developed as follows:

$$[P(\psi)]\frac{\mathrm{d}\left([P(\psi)]^{-1}\right)}{\mathrm{d}t} = \begin{bmatrix} 0 & -1 & 0 \\ 1 & 0 & 0 \\ 0 & 0 & 0 \end{bmatrix}\frac{\mathrm{d}\psi}{\mathrm{d}t} \tag{2.93}$$

The formula of the differential equations is:

$$\left\{\begin{aligned} \frac{\mathrm{d}\phi_{\mathrm{sd}}}{\mathrm{d}_t} &= v_{\mathrm{sd}} - R_s i_{\mathrm{sd}} + \omega_s\phi_{\mathrm{sq}} \\ \frac{\mathrm{d}\phi_{\mathrm{sq}}}{\mathrm{d}_t} &= v_{\mathrm{sq}} - R_s i_{\mathrm{sq}} - \omega_s\phi_{\mathrm{sd}} \\ \frac{\mathrm{d}\phi_{s0}}{\mathrm{d}_t} &= v_{s0} - R_s i_{s0} \end{aligned}\right\} \tag{2.94}$$

$$\left\{\begin{aligned} \frac{\mathrm{d}\phi_{\mathrm{rd}}}{\mathrm{d}_t} &= v_{\mathrm{rd}} - R_r i_{\mathrm{rd}} + \omega_r\phi_{\mathrm{rq}} \\ \frac{\mathrm{d}\phi_{\mathrm{rq}}}{\mathrm{d}_t} &= v_{\mathrm{rq}} - R_r i_{\mathrm{rq}} - \omega_r\phi_{\mathrm{rd}} \\ \frac{\mathrm{d}\phi_{r0}}{\mathrm{d}_t} &= v_{r0} - R_r i_{r0} \end{aligned}\right\} \tag{2.95}$$

2.4.2.3 Determination of Equations Between Flux and Currents

According to the relations existing between flux and currents (Eq. 2.28), the Park reference frame coordinates are expressed with the inverse Park transformation:

$$\begin{bmatrix} [P(\theta_s)]^{-1}[\phi_{\mathrm{s\_dq0}}] \\ [P(\theta_r)]^{-1}[\phi_{\mathrm{r\_dq0}}] \end{bmatrix} = \begin{bmatrix} [L_s] & [M_{\mathrm{sr}}(\theta)] \\ [M_{\mathrm{sr}}(\theta)] & [L_r] \end{bmatrix} \begin{bmatrix} [P(\theta_s)]^{-1}[I_{\mathrm{s\_dq0}}] \\ [P(\theta_r)]^{-1}[I_{\mathrm{r\_dq0}}] \end{bmatrix} \tag{2.96}$$

Multiplying on the left by the Park transformation, results in:

$$\begin{bmatrix} [\phi_{\mathrm{s\_dq0}}] \\ [\phi_{\mathrm{r\_dq0}}] \end{bmatrix} = \begin{bmatrix} [P(\theta_s)][L_s][P(\theta_s)]^T & [P(\theta_s)][M_{\mathrm{sr}}(\theta)][P(\theta_r)]^T \\ [P(\theta_r)][M_{\mathrm{sr}}(\theta)]^T[P(\theta_s)]^T & [P(\theta_r)][L_r][P(\theta_r)]^T \end{bmatrix} \begin{bmatrix} [I_{\mathrm{s\_dq0}}] \\ [I_{\mathrm{r\_dq0}}] \end{bmatrix} \tag{2.97}$$

Developing the previous formula leads to the expression of four submatrices that define the couplings of the equivalent model in the two-axe frame:

$$\begin{bmatrix} [\phi_{\mathrm{s\_dq0}}] \\ [\phi_{\mathrm{r\_dq0}}] \end{bmatrix} = \begin{bmatrix} [L_{\mathrm{ps}}] & [M_{\mathrm{psr}}] \\ [M_{\mathrm{psr}}]^{\mathrm{T}} & [L_{\mathrm{pr}}] \end{bmatrix} \begin{bmatrix} [I_{\mathrm{s\_dq0}}] \\ [I_{\mathrm{r\_dq0}}] \end{bmatrix} \tag{2.98}$$

with

$$[L_{\mathrm{ps}}] = \begin{bmatrix} L_s & 0 & 0 \\ 0 & L_s & 0 \\ 0 & 0 & L_{s0} \end{bmatrix} \tag{2.99}$$

$$[L_{\mathrm{pr}}] = \begin{bmatrix} L_r & 0 & 0 \\ 0 & L_r & 0 \\ 0 & 0 & L_{r0} \end{bmatrix} \tag{2.100}$$

$$[M_{\mathrm{psr}}] = [P(\theta_s)]^{-1}[M_{\mathrm{sr}}(\theta)][P(\theta_r)] = \frac{3}{2}\, M_{\mathrm{sr}}[R_r] \tag{2.101}$$

$$[R_r] = \begin{bmatrix} 1 & 0 & 0 \\ 0 & 1 & 0 \\ 0 & 0 & 0 \end{bmatrix} \tag{2.102}$$

Using the Park transformation, it can be noted that the mutual inductances of Eq. 2.98 do not vary with the rotor position. A model with constant coefficients is then obtained. The number of parameters to use is significantly less. Furthermore, as the matrices are now diagonal, the equations referring respectively to axes d and q are not coupled.

2.4.2.4 Torque Calculation

Starting from the torque expression in the three-axe frame (2.38), applying the Park transform to the current results in:

$$T = p\left([P(\theta)]^{-1}[I_{\mathrm{s\_dq0}}]\right)^{\mathrm{T}} \frac{d[M_{\mathrm{sr}}(\theta)]}{d\theta} [P(\theta)]^{-1}[I_{\mathrm{r\_dq0}}] \tag{2.103}$$

Table 2.2 Causal organization of the mathematical equations in the Park reference frame

Electromagnetic part	
$R1_{sd}$: $e_{\phi_{sd}} = v_{sd} - v_{Rsd} - e_{sd}$	$R1_{rd}$: $e_{\phi_{rd}} = v_{rd} - v_{Rrd} - e_{rd}$
$R1_{sq}$: $e_{\phi sq} = v_{sq} - v_{Rsq} + e_{sq}$	$R1_{rq}$: $e_{\phi rq} = v_{rq} - v_{Rrq} + e_{rq}$
$R2_{sd}$: $\phi_{sd} = \int_{t_0}^{t_0+\Delta t} e_{\phi_{sd}}\, dt + \phi_{sd}(t_0)$	$R2_{rd}$: $\phi_{rd} = \int_{t_0}^{t_0+\Delta t} e_{\phi_{rd}}\, dt + \phi_{rd}(t_0)$
$R2_{sq}$: $\phi_{sq} = \int_{t_0}^{t_0+\Delta t} e_{\phi_{sq}}\, dt + \phi_{sq}(t_0)$	$R2_{rq}$: $\phi_{rq} = \int_{t_0}^{t_0+\Delta t} e_{\phi_{rq}}\, dt + \phi_{rq}(t_0)$
$R3_{sd}$: $i_{sd} = \frac{1}{L_s}(\phi_{sd} - Mi_{rd})$	$R3_{rd}$: $i_{rd} = \frac{1}{L_r}(\phi_{rd} - Mi_{sd})$
$R3_{sq}$: $i_{sq} = \frac{1}{L_s}(\phi_{sq} - Mi_{rq})$	$R3_{rq}$: $i_{rq} = \frac{1}{L_r}(\phi_{rq} - Mi_{sq})$
$R4_{sd}$: $v_{Rsd} = R_s i_{sd}$	$R4_{rd}$: $v_{Rrd} = R_r i_{rd}$
$R4_{sq}$: $v_{Rsq} = R_s i_{sq}$	$R4_{rq}$: $v_{Rrq} = R_r i_{rq}$
Electromagnetic coupling:	
$R5_{sd}$: $T_{sd} = -p\, i_{sd}\, \phi_{sq}$	$R5_{rd}$: $T_{rd} = -p\frac{M}{L_r}\, i_{sd}\, \phi_{rq}$
$R5_{sq}$: $T_{sq} = p\, i_{sq}\, \phi_{sd}$	$R5_{rq}$: $T_{rq} = p\frac{M}{L_r}\, i_{sq}\phi_{rd}$
$R6_{sd}$: $e_{sd} = -\phi_{sq}\, \omega s$	$R6_{rd}$: $e_{rd} = -\phi_{rq}\, \omega r$
$R6_{sq}$: $e_{sq} = -\phi_{sd}\omega s$	$R6_{rq}$: $e_{rq} = -\phi_{rd}\omega r$
$R7_s$: $T_s = T_{sd} + T_{sq}$	$R7_r$: $T_r = T_{rd} + T_{rq}$
Speed of the reference frame : $R8$: $\omega_s = \omega_r + p$	

after introducing (2.101), the results are:

$$T = p\left[I_{s\_dq0}\right]^{T}\frac{3}{2}M_{sr}\frac{d\left[R_p\right]}{d\theta}\left[I_{r\_dq0}\right] \tag{2.104}$$

and using Eq. 2.73, the result in scalar form is:

$$T = pM\left(i_{sq}i_{rd} - i_{sd}i_{rq}\right) \tag{2.105}$$

When using the equations that link flux to currents (Eq. 2.98), it is possible to determine other torque equivalent expressions more adapted to the conception of the squirrel-cage machine control. Among the ones that are used most, there is the equation in the rotor flux frame, which brings into play the (measurable) currents to the stator and the flows to the rotor. Based on the expressions of the rotor currents (relations $R3_{rd}$ and $R3_{rq}$, Table 2.2), it results in:

$$T = p\frac{M}{L_r}\left(i_{sq}\phi_{rd} - i_{sd}\,\phi_{rq}\right) \tag{2.106}$$

Similarly, by substituting the expressions of the rotor currents by relations R3sd, R3sq (Table 2.2) in Eq. 2.105, it is possible to determine an expression of the torque, which uses the stator flux components:

$$T = p\left(i_{sq}\phi_{sd} - i_{sd}\phi_{sq}\right) \tag{2.107}$$

2.4.3 General Model in the Park Reference Frame

2.4.3.1 Setting in Equation of the Electromagnetical Part

Using the mesh law, Eqs. 2.94 and 2.95 are rewritten to make virtual induction e.m.f. corresponding to axes d ($R1_{\mathrm{sd}}$) and ($R1_{\mathrm{sq}}$), and to axes q ($R1_{\mathrm{rd}}$) and ($R1_{\mathrm{rq}}$) appear (Table 2.2).

This formulation uses virtual voltages across the terminals of resistances corresponding to axes d [($R4_{\mathrm{sd}}$) and ($R4_{\mathrm{rd}}$)] and to axes q [($R4_{\mathrm{sq}}$) and ($R4_{\mathrm{rq}}$)] along with electromotive rotation forces relative to axes d [($R6_{\mathrm{sd}}$) and ($R6_{\mathrm{rd}}$)] and to axes q [($R6_{\mathrm{sq}}$) and ($R6_{\mathrm{rq}}$)]. Integrating these induction e.m.f. allows the calculation of the various components of the fluxes. It is then possible to determine the currents by using the inverse equations of coupling with the fluxes (relation $(2.98)^{-1}$), which can be expressed by:

$$\begin{bmatrix} i_{\mathrm{sq}} \\ i_{\mathrm{rq}} \end{bmatrix} = \frac{1}{L_r L_s - M^2} \begin{bmatrix} L_r & -M \\ -M & L_s \end{bmatrix} \begin{bmatrix} \phi_{\mathrm{sq}} \\ \phi_{\mathrm{rq}} \end{bmatrix} \tag{2.108}$$

$$\begin{bmatrix} i_{\mathrm{sd}} \\ i_{\mathrm{rd}} \end{bmatrix} = \frac{1}{L_r L_s - M^2} \begin{bmatrix} L_r & -M \\ -M & L_s \end{bmatrix} \begin{bmatrix} \phi_{\mathrm{sd}} \\ \phi_{\mathrm{rd}} \end{bmatrix} \tag{2.109}$$

2.4.3.2 Equivalent Park Machine

v_{sd} and v_{sq} are the voltages applied to the windings of a fictive stator, the magnetic axes of which would be axes d and q that produce the same magnetic effects as obtained with true windings. v_{rd} and v_{rq} are the voltages applied to the windings of a fictive rotor, the magnetic axes of which would be axes d and q producing the same magnetic effects as obtained with true windings.

Thus, the Park transform allows the substitution of the real three-phase system by another (see Fig. 2.9) composed of:

- two stator (of resistance R_s and inductance L_s) and rotor (of resistance R_r and inductance L_r) windings rotating at an angular speed ω_s and run through by direct and quadrature currents. The frame of reference $\left(\vec{O}_{\mathrm{sa}}, \vec{O}_{\mathrm{sb}}, \vec{O}_{\mathrm{sc}}\right)$ is stationary.
- two stationary R-L circuits, run through by homopolar currents (i_{r0} and i_{s0}).

A coupling remains between the rotor and stator windings which are located on the same axis. The coupling is expressed by the mutual inductance M.

It is demonstrated that if the machine is supplied by a balanced three-phase system, a field rotates in the air gap at an angular speed, which is equal to the speed of the pulsation of the voltages, and therefore equal to the currents, as well [Caron and Hautier 1995]. When choosing a frame (d, q) linked to the rotating

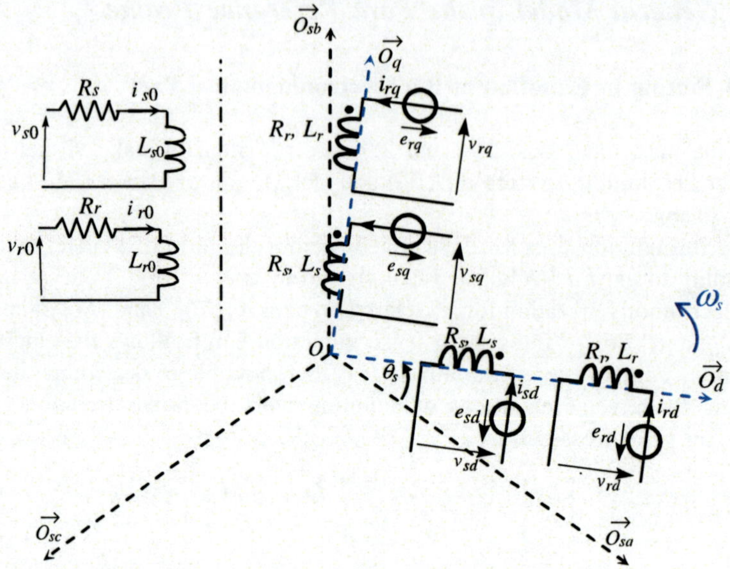

Fig. 2.9 Equivalent two-phase windings rotating in the park frame

field and then rotating at an angular speed, which is equal to the speed of the voltages pulsation supplying the true machine, the equivalent system in that frame creates a similar magnetic field. Quantities v_{sd} and $v_{\mathrm{sq}}$are then continuous voltages under steady conditions. The stationary three-phase windings—run through by sinusoidal currents—are then substituted by rotating two-phase windings—run through by direct currents. Frame (d, q)—linked to rotor quantities—then has to turn at the same speed for the mutual inductances to be constant. Condition (2.85) has to be checked to obtain this. The rotor and the stator of the machine—then called « Park machine » —are rotating at the same speed, so that flux and currents are linked by an independent time expression.

2.4.4 Model Using the Rotor Flux

2.4.4.1 Causal Ordering Graph Principle and Model

Several equivalent expressions can be used for the calculation of the torque. The expression using the flux to the rotor (Eq. 2.106) can be divided into two parts:

$$R5_{\mathrm{rq}} \rightarrow T_{\mathrm{rq}} = p\frac{M}{L_r}\, i_{\mathrm{sq}}\, \phi_{\mathrm{rd}} \tag{2.110}$$

and

$$R5_{\text{rd}} \rightarrow T_{\text{rd}} = -p\frac{M}{L_r} i_{\text{sd}}\, \phi_{\text{rq}} \tag{2.111}$$

The expression is then written as follows:

$$R7_s \rightarrow T = T_{\text{rd}} + T_{\text{rq}} \tag{2.112}$$

This expression is used for the implementation of the control known as "rotor flux-oriented".

The COG of this model (Fig. 2.10) emphasizes two gyrator couplings [relations ($R5_{\text{rd}}$), ($R5_{\text{rq}}$), ($R6_{\text{rd}}$) and ($R6_{\text{rq}}$)], since the expression of the electromagnetic torque shows that the torque is composed of two components called c_{rd} (for axis *d*) andc_{rq} (for axis *q*). The Park machine can also be considered as the association of two fictive direct current machines that are mechanically and electrically coupled. This mathematical model can be considered as a system of which the input quantities are a stator voltage vector and the mechanical speed imposed by the mechanical load (see paragraph 1.4.2.4.). The output quantities of this model are the generated electromechanical torque and the vector of the stator currents. A global representation of a squirrel-cage machine can easily be obtained by canceling the rotor voltages.

The Park transform allows the calculation of the direct and quadrature components of the voltages applied to the machine. The transform inverts the currents in the three-axe frame. The orientation angle of these two transforms is calculated by integrating the pulsation of the frame linked to the stator, with a reset when the stator voltage of the phase is zero, at:

$$\theta_s = \int_{t_0}^{t_0+\Delta t} \omega_s \, \text{dt} + \theta_s(t_0) \tag{2.113}$$

Figure 2.11 shows the choice of linking the Park reference frame to the pulsation of the stator quantities. Consequently, the pulsation of these quantities is null in that frame. So, the Park transform is a mathematical operator which allows the conversion of sinusoidal quantities of pulsation $\omega_s$in a quantity of null value. This transform then allows the determination of a medium equivalent model for quantities with a given pulsation or with a pulsation more or less equal to it.

2.4.4.2 State Space Representation

In equations ($R2_{\text{sd}}, R2_{\text{sq}}, R2_{\text{rd}}, R2_{\text{rq}}$), the different components of the stator and rotor fluxes are obtained by integration—they are also called state variables.

As the fluxes and currents are linked by expressions ($R3_{\text{sd}}, R3_{\text{sq}}, R3_{\text{rd}}, R3_{\text{rq}}$), the different current components, or even a combination of fluxes and currents, can be chosen as state variables. For the conception of a motor control, it can be

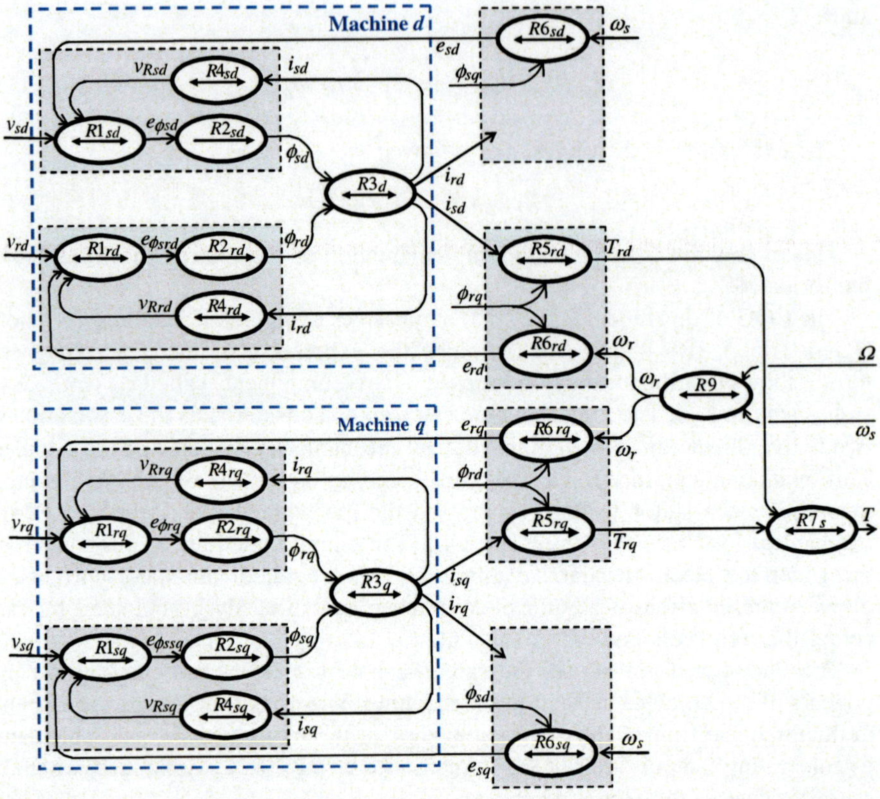

Fig. 2.10 C.O.G. of Park's model using fluxes to the rotor (the relations are carried over in Table 2.2)

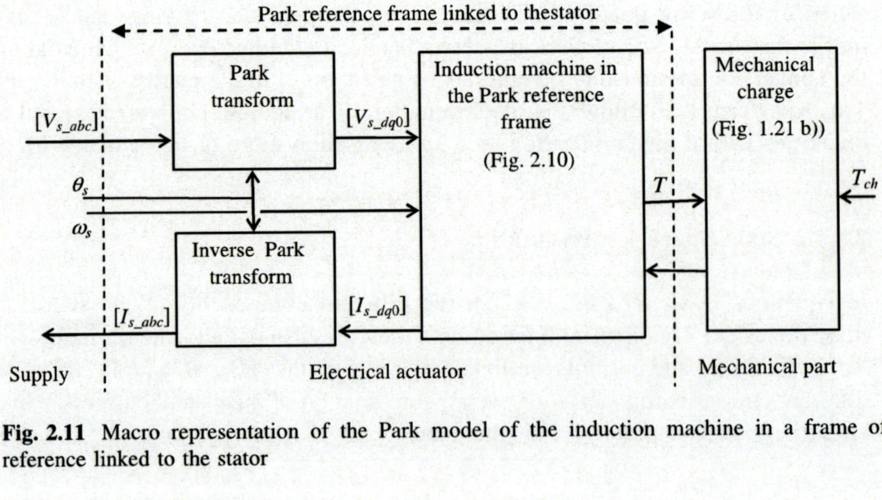

Fig. 2.11 Macro representation of the Park model of the induction machine in a frame of reference linked to the stator

interesting to choose the stator current components for state variables. In fact, this current as well as the rotor flux components—in as much as this flux will be controlled—can be easily measured. The state space representation then consists of determining only the differential equations that allow the definition of these variables by integration with intermediary mathematical relations. Thus, relations $R1_{\text{rd}}, R4_{\text{rd}}$ and $R6_{\text{rd}}$ are injected in the derivative of this flux $(R2_{\text{rd}})$ as to determine the differential equation of the rotor flux direct component. When applying the same procedure to other fluxes, this results in:

$$\frac{\mathrm{d}\phi_{\text{sd}}}{\mathrm{d}t} = v_{\text{sd}} - R_s i_{\text{sd}} + \omega_s \phi_{\text{sq}} \tag{2.114a}$$

$$\frac{\mathrm{d}\phi_{\text{sq}}}{\mathrm{d}t} = v_{\text{sq}} - R_s i_{\text{sq}} - \omega_s \phi_{\text{sd}} \tag{2.114b}$$

$$\frac{\mathrm{d}\phi_{\text{rd}}}{\mathrm{d}t} = v_{\text{rd}} - R_r i_{\text{rd}} + \omega_r \phi_{\text{rq}} \tag{2.114c}$$

$$\frac{\mathrm{d}\phi_{\text{rq}}}{\mathrm{d}t} = v_{\text{rq}} - R_r i_{\text{rq}} - \omega_r \phi_{\text{rd}} \tag{2.114d}$$

When introducing relations $R3_{\text{rd}}$ and $R3_{\text{rq}}$ in $R3_{\text{sd}}$ and $R3_{\text{sq}}$, and defining the dispersion coefficient:

$$\sigma = 1 - \frac{M^2}{L_r L_s}, \tag{2.115}$$

This results in:

$$\phi_{\text{sd}} = \sigma L_s i_{\text{sd}} + \frac{M}{L_r} \phi_{\text{rd}} \tag{2.116a}$$

$$\phi_{\text{sq}} = \sigma L_s\, i_{\text{sq}} + \frac{M}{L_r} \phi_{\text{rq}} \tag{2.116b}$$

Finally, the following equations are obtained when replacing the stator flux components by (2.116a) and the rotor flux components by $R3_{\text{rd}}$ and $R3_{\text{rq}}$ in the differential equations of the fluxes (2.114a):

$$\sigma L_s \frac{\mathrm{d}i_{\text{sd}}}{\mathrm{d}t} = v_{\text{sd}} - R_s i_{\text{sd}} + \omega_s \sigma L_s i_{\text{sq}} - \frac{M}{L_r}\frac{\mathrm{d}\phi_{\text{rd}}}{\mathrm{d}t} + \omega_s \frac{M}{L_r} \phi_{\text{rq}} \tag{2.117a}$$

$$\sigma L_s \frac{\mathrm{d}i_{\text{sq}}}{\mathrm{d}t} = v_{\text{sq}} - R_s i_{\text{sq}} - \omega_s \sigma L_s i_{\text{sd}} - \frac{M}{L_r}\frac{\mathrm{d}\phi_{\text{rq}}}{\mathrm{d}t} - \omega_s \frac{M}{L_r} \phi_{\text{rd}} \tag{2.117b}$$

$$\frac{\mathrm{d}\phi_{\text{rd}}}{\mathrm{d}t} = v_{\text{rd}} - \frac{R_r}{L_r} \phi_{\text{rd}} + \frac{R_r}{L_r} M i_{\text{sd}} + \omega_r \phi_{\text{rq}} \tag{2.117c}$$

$$\frac{d\phi_{rq}}{dt} = v_{rq} - \frac{R_r}{L_r}\phi_{rq} + \frac{R_r}{L_r}Mi_{sq} - \omega_r\phi_{rd} \tag{2.117d}$$

In the two last equations the derivatives of the rotor fluxes appear, which can be replaced by their expression (2.117c) and (2.117d):

$$\sigma L_s\frac{di_{sd}}{dt} = v_{sd} - \frac{M}{L_r}v_{rd} - R_{sr}i_{sd} + \omega_s\sigma L_s i_{sq} + \frac{M}{L_r}(\omega_s - \omega_r)\,\phi_{rq} + \frac{M}{L_r^2}R_r\phi_{rd} \tag{2.118a}$$

$$\sigma L_s\frac{di_{sq}}{dt} = v_{sq} - \frac{M}{L_r}v_{rq} - R_{sr}i_{sq} - \omega_s\sigma L_s i_{sd} - \frac{M}{L_r}(\omega_s - \omega_r)\,\phi_{rd} + \frac{M}{L_r^2}R_r\phi_{rq} \tag{2.118b}$$

$$\frac{d\phi_{rd}}{dt} = v_{rd} - \frac{R_r}{L_r}\phi_{rd} + \frac{R_r}{L_r}M\,i_{sd} + \omega_r\phi_{rq} \tag{2.119a}$$

$$\frac{d\phi_{rq}}{dt} = v_{rq} - \frac{R_r}{L_r}\phi_{rq} + \frac{R_r}{L_r}M\,i_{sq} - \omega_r\phi_{rd} \tag{2.119b}$$

with R_{sr}, the total resistance viewed from the stator is:

$$R_{sr} = R_s + R_r\frac{M^2}{L_r^2} = R_s + \frac{(1-\sigma)L_s}{L_r}R_r \tag{2.120}$$

When using Eq. 2.86, these four relations are expressed in vector form as follows:

$$\begin{bmatrix} \dot{\phi}_{rd} \\ \dot{\phi}_{rq} \\ \dot{i}_{sd} \\ \dot{i}_{sq} \end{bmatrix} = \begin{bmatrix} -\frac{R_r}{L_r} & \omega_r & M\frac{R_r}{L_r} & 0 \\ -\omega_r & -\frac{R_r}{L_r} & 0 & M\frac{R_r}{L_r} \\ \frac{M R_r}{\sigma L_s L_r^2} & \frac{M}{\sigma L_s L_r}p\Omega & -\frac{R_{sr}}{\sigma L_s} & \omega_s \\ \frac{M}{\sigma L_s L_r}p\Omega & \frac{M R_r}{\sigma L_s L_r^2} & -\omega_s & -\frac{R_{sr}}{\sigma L_s} \end{bmatrix} \begin{bmatrix} \phi_{rd} \\ \phi_{rq} \\ i_{sd} \\ i_{sq} \end{bmatrix} + \begin{bmatrix} 1 & 0 & 0 & 0 \\ 0 & 1 & 0 & 0 \\ \frac{-M}{\sigma L_r L_s} & 0 & \frac{1}{\sigma L_s} & 0 \\ 0 & \frac{-M}{\sigma L_r L_s} & 0 & \frac{1}{\sigma L_s} \end{bmatrix} \begin{bmatrix} v_{rd} \\ v_{rq} \\ v_{sd} \\ v_{sq} \end{bmatrix} \tag{2.121}$$

This mathematical formulation of the model will be used to elaborate the rotor-flux-oriented control without a priori knowledge of the squirrel-cage induction machine in Chap. 3.

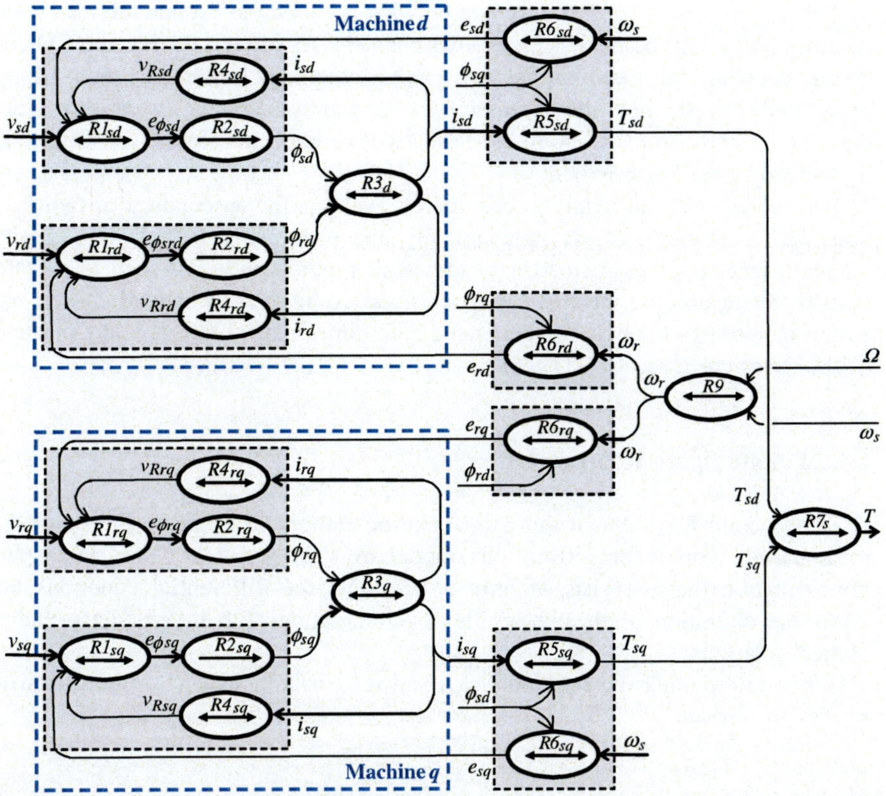

Fig. 2.12 C.O.G. of Park's model using fluxes to the stator (the relations are reported in Table 2.2)

2.4.5 Model Using the Stator Flux

2.4.5.1 Model Principle and Causal Ordering Graph

In the stator flux frame of reference, the torque expression that uses fluxes to the stator (Eq. 2.107) can be divided into two parts:

$$R5_{\mathrm{sd}} \rightarrow T_{\mathrm{sq}} = p i_{\mathrm{sq}} \phi_{\mathrm{sd}} \tag{2.122}$$

and

$$R5_{\mathrm{sq}} \rightarrow T_{\mathrm{sd}} = -p i_{\mathrm{sd}} \phi_{\mathrm{sq}} \tag{2.123}$$

It is described as follows:

$$R7_{s} \rightarrow T = T_{\mathrm{sd}} + T_{\mathrm{sq}} \tag{2.124}$$

This expression is used to elaborate on what is called the "stator-flux-oriented" control (which will be dealt with in next chapter). As proven earlier, the COG of this model (Fig. 2.12) emphasizes two gyrator couplings [relations ($R5_{sd}$), ($R5_{sq}$), ($R6_{sd}$) and ($R6_{sq}$)], since the expression of the electromagnetic torque shows that the torque expression is composed of two components called T_{sd} and T_{sq}, respectively axis d and axis q.

Park's machine can again be considered again as the association of two fictitious DC machines that are mechanically and electrically linked. This mathematical model is equivalent to a system with input quantities—such as a stator voltage vector and the mechanical speed imposed by the mechanical charge, and output quantities—such as the generated electromechanical torque and the vector of the stator currents.

2.4.5.2 State Space Representation

A second control system of this motor can be elaborated by using the stator and rotor current components from this model as state variables. The state space representation then consists of only determining the differential equations that allow the definition of these variables by integration with intermediary mathematical relations.

When introducing the relations $R3_{sd}$ and $R3_{sq}$ in $R3_{rd}$ and $R3_{rq}$, the following results are shown:

$$\phi_{rd} = \sigma L_r i_{rd} + \frac{M}{L_s}\phi_{sd} \tag{2.125a}$$

$$\phi_{rq} = \sigma L_r i_{rq} + \frac{M}{L_s}\phi_{sq} \tag{2.125b}$$

When replacing the stator flux components by (2.125a) and the stator current components by $R3_{sd}$ and $R3_{sq_{in}}$ the flux differential equations are obtained: (2.114a):

$$\frac{d\phi_{sd}}{dt} = v_{sd} - \frac{R_s}{L_s}\phi_{sd} + \frac{R_s}{L_s}M i_{rd} + \omega_s\phi_{sq} \tag{2.126a}$$

$$\frac{d\phi_{sq}}{dt} = v_{sq} - \frac{R_s}{L_s}\phi_{sq} + \frac{R_s}{L_s}M i_{rq} - \omega_s\phi_{sd} \tag{2.126b}$$

$$\sigma L_r\frac{di_{rd}}{dt} = v_{rd} - R_r i_{rd} + \omega_r\sigma L_r i_{rq} - \frac{M}{L_s}\frac{d\phi_{sd}}{dt} + \omega_r\frac{M}{L_s}\phi_{sq} \tag{2.126c}$$

$$\sigma L_r\frac{di_{rq}}{dt} = v_{rq} - R_r i_{rq} - \omega_r\sigma L_r i_{rd} - \frac{M}{L_s}\frac{d\phi_{sq}}{dt} - \omega_r\frac{M}{L_s}\phi_{sd} \tag{2.126d}$$

The last two equations show the derivatives of the stator fluxes that can be replaced by their expressions (2.126a) and (2.126b)

$$\frac{d\phi_{sd}}{dt} = v_{sd} - \frac{R_s}{L_s}\phi_{sd} + \frac{R_s}{L_s}Mi_{rd} + \omega_s\phi_{sq} \tag{2.127a}$$

$$\frac{d\phi_{sq}}{dt} = v_{sq} - \frac{R_s}{L_s}\phi_{sq} + \frac{R_s}{L_s}Mi_{rq} - \omega_s\phi_{sd} \tag{2.127b}$$

$$\sigma L_r\frac{di_{rd}}{dt} = v_{rd} - \frac{M}{L_s}v_{sd} - R_{rs}i_{rd} + \omega_r\sigma L_r i_{rq} + \frac{M}{L_s}(\omega_r - \omega_s)\phi_{sq} + \frac{M}{L_s^2}R_s\phi_{sd} \tag{2.127c}$$

$$\sigma L_r\frac{di_{rq}}{dt} = v_{rq} - \frac{M}{L_s}v_{sq} + R_{rs}i_{rq} - \omega_r\sigma L_r i_{rd} - \frac{M}{L_s}(\omega_s - \omega_r)\phi_{sd} + \frac{M}{L_s^2}R_s\phi_{sq} \tag{2.127d}$$

with R_{rs}, the total resistance shown by the rotor is:

$$R_{rs} = R_r + R_s\frac{M^2}{L_s^2} = R_r + \frac{(1-\sigma)L_r}{L_s}R_s \tag{2.128}$$

These four relations are shown in matrix form as follows:

$$\begin{bmatrix} \dot{\phi}_{sd} \\ \dot{\phi}_{sq} \\ \dot{i}_{rd} \\ \dot{i}_{rq} \end{bmatrix} = \begin{bmatrix} -\frac{R_s}{L_s} & \omega_s & M\frac{R_s}{L_s} & 0 \\ -\omega_s & -\frac{R_s}{L_s} & 0 & M\frac{R_s}{L_s} \\ \frac{M R_s}{L_s^2\sigma L_r} & \frac{M}{L_r\sigma L_s}(\omega_r - \omega_s) & \frac{-R_{rs}}{\sigma L_r} & \omega_r \\ \frac{M}{L_r\sigma L_s}(\omega_s - \omega_r) & \frac{M R_s}{L_s^2\sigma L_r} & -\omega_r & \frac{-R_{rs}}{\sigma L_r} \end{bmatrix} \begin{bmatrix} \phi_{sd} \\ \phi_{sq} \\ i_{rd} \\ i_{rq} \end{bmatrix} + \begin{bmatrix} 1 & 0 & 0 & 0 \\ 0 & 1 & 0 & 0 \\ \frac{-M}{\sigma L_r L_s} & 0 & \frac{1}{\sigma L_r} & 0 \\ 0 & \frac{-M}{\sigma L_r L_s} & 0 & \frac{1}{\sigma L_r} \end{bmatrix} \begin{bmatrix} v_{sd} \\ v_{sq} \\ v_{rd} \\ v_{rq} \end{bmatrix} \tag{2.129}$$

This mathematical formulation of the model is used to elaborate on the stator-flux-oriented control of the squirrel-cage induction machine [Degobert 1997], and for the direct torque control [Canudas de Wit 2000].

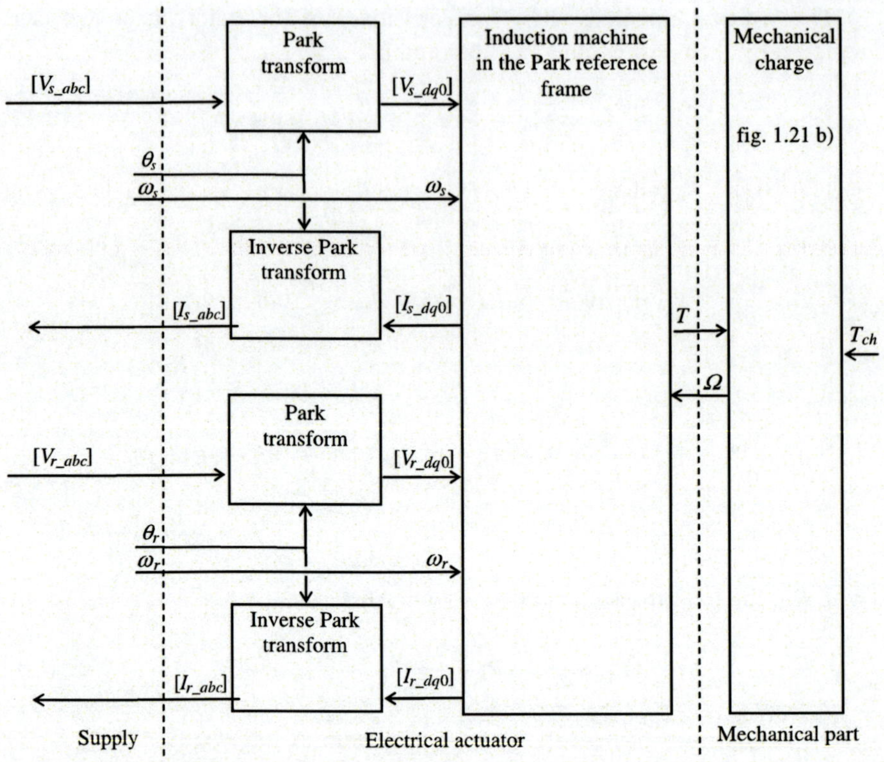

Fig. 2.13 Representation of the induction machine model into a system

2.4.6 Model for System Analysis

The previously developed models show that the induction machine is an electromechanical converter that:

- receives voltages to the stator and rotor as electrical quantities and relieves currents;
- receives speed as a mechanical quantity and that relieves an electromotive torque.

The second comment assumes that the mechanical part of the machine (usually negligible compared to the mechanical part of the load) is modeled externally with the mechanical part of the charge.

Figure 2.13 also illustrates the fact that the Park transform is a mathematical operator that allows converting sinusoidal quantities of pulsation ω_s into a null value quantity.

Thus, this transformation allows the determination of a similar average model for quantities that have a specific pulsation or that are more or less equal to this one.

It is useless to use a model that is neither based on the physical quantities internal to the machine (flux, e.m.f., etc.) nor on their coupling in order to carry out studies on industrial devices using this machine technology. It is, however, useful to have a model that explicits and especially considers the physical quantities exchanged between the machine and the external elements (power converters, mechanical charge, etc.) associated with it. As far as the electrical part of the machine is concerned, it is then necessary to determine the currents according to the applied voltages by replacing the fluxes in the Eq. 2.114a with their expressions ($R3_{sd}$, $R3_{sq}$, $R3_{rd}$, $R3_{rq}$) (Table 2.2). The direct axis components are expressed as follows:

$$L_s \frac{di_{sd}}{dt} + M \frac{di_{rd}}{dt} = v_{sd} - R_s i_{sd} + \omega_s (L_s i_{sq} + M i_{rq}) \tag{2.130}$$

$$L_r \frac{di_{rd}}{dt} + M \frac{di_{sd}}{dt} = v_{rd} - R_r i_{rd} + \omega_r (L_r i_{rq} + M i_{sq}) \tag{2.131}$$

When the expression of the derivative of the stator current direct component from the second equation is replaced in the first equation, the previous device is then expressed:

$$\frac{di_{sd}}{dt} = \frac{1}{\sigma L_s} \left[v_{sd} - R_s i_{sd} - \frac{M}{L_r} v_{rd} - e'_{sd} \right] \tag{2.132a}$$

$$\frac{di_{rd}}{dt} = \frac{1}{\sigma L_r} \left[v_{rd} - R_r i_{rd} - \frac{M}{L_s} v_{sd} - e'_{rd} \right] \tag{2.132b}$$

with the following coupling e.m.f., the following results are shown:

$$e'_{sd} = -M \frac{R_r}{L_r} i_{rd} - \omega_s (L_s i_{sq} + M i_{rq}) + \omega_r \frac{M}{L_r} (L_r i_{rq} + M i_{sq}) \tag{2.133a}$$

$$e'_{rd} = -M \frac{R_s}{L_s} i_{sd} - \omega_r (L_r i_{rq} + M i_{sq}) + \omega_s \frac{M}{L_r} (L_s i_{sq} + M i_{rq}) \tag{2.133b}$$

The same can be said of the quadrature components:

$$\frac{di_{sq}}{dt} = \frac{1}{\sigma L_s} \left[v_{sq} - R_s i_{sq} - \frac{M}{L_r} v_{rq} - e'_{sq} \right] \tag{2.134a}$$

$$\frac{di_{rq}}{dt} = \frac{1}{\sigma L_r} \left[v_{rq} - R_r i_{rq} - \frac{M}{L_s} v_{sq} - e'_{rq} \right] \tag{2.134b}$$

with the following coupling e.m.fs:

$$e'_{sq} = -M \frac{R_r}{L_r} i_{rq} + \omega_s (L_s i_{sd} + M i_{rd}) - \omega_r \frac{M}{L_r} (L_r i_{rd} + M i_{sd}) \tag{2.135a}$$

$$e'_{rq} = -M\frac{R_s}{L_s}i_{sq} + \omega_r(L_r i_{rd} + M i_{sd}) - \omega_s\frac{M}{L_r}(L_s i_{sd} + M i_{rd}) \qquad (2.135b)$$

Therefore, the electromotive torque can be calculated with the Eq. 2.105.

2.5 The Effect of the Magnetic Saturation

Previously, the dynamic models have been developed in the assumption that the machine does not suffer from magnetic saturation (Sect. 2.2.1), which consists in considering a linear characteristic $B = f(H)$ (Fig. 1.5), or even a linear characteristic between flux and current. In Chap. 3, control systems are developed from these models. In order to validate these controls and to study their sensibility to the real presence of these saturations, it is necessary that a simple model of this magnetic saturation is introduced, which is valid under steady conditions. Thus, a model that makes a linear parameter β interfere is chosen to take into account the air gap, and another non linear parameter (elevation to the power s) to broach the specificity of the saturation point (so appears from experimental results of Fig. 2.15 for $i_{mn} > 0,7$) :

$$i_{mn} = \beta\phi_{sn} + (1-\beta)\phi_{sn}^{s} \qquad (2.136)$$

$$M_n = \frac{\phi_{sn}}{i_{mn}} \qquad (2.137)$$

i_{mn}, ϕ_{sn} and M_n are respectively the standardized values compared to their rated value (also called "per unit"):

- of the magnetizing current

$$i_m = \sqrt{(i_{sd} + i_{rd})^2 + (i_{sq} + i_{rq})^2} \qquad (2.138)$$

- of the stator flux

$$\phi_s = \sqrt{\phi_{sd}^2 + \phi_{sq}^2} \qquad (2.139)$$

- and of static inductance $\bar{M}$.

The organization of all equations leads to the calculation of the static inductance (Fig. 2.14).

The dynamic inductance does not necessarily have to be taken into account as the sensibility study will be carried out under steady conditions [Vas 1990].

The experimental and theoretical magnetic characteristics are compared with one another in order to determine the parameters [β and s of the expression (2.136)].

Therefore, a no-load test is carried out to determine the magnetic characteristic $U_s = f(I_m)$, with U_s representing the effective stator voltage and I_M representing

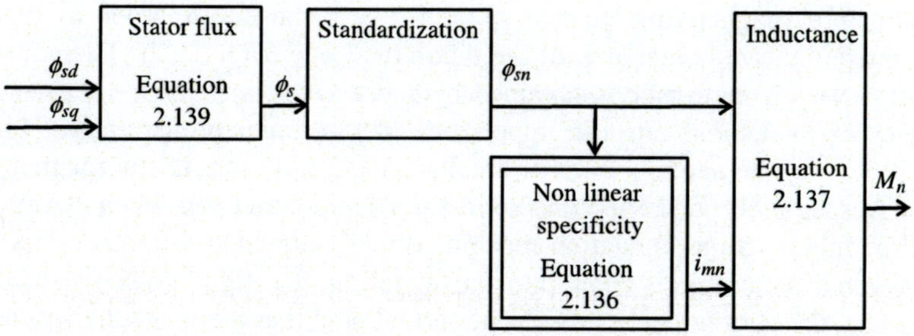

Fig. 2.14 Representation of the induction machine model into a device

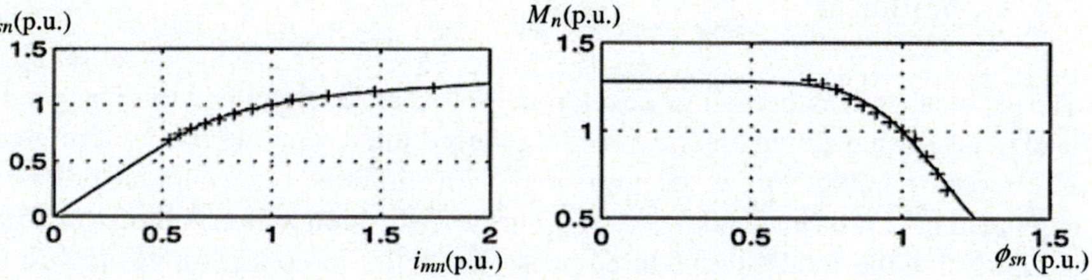

Fig. 2.15 Comparison between the experimental data (*dashed* line) and the saturation model (*solid* line)

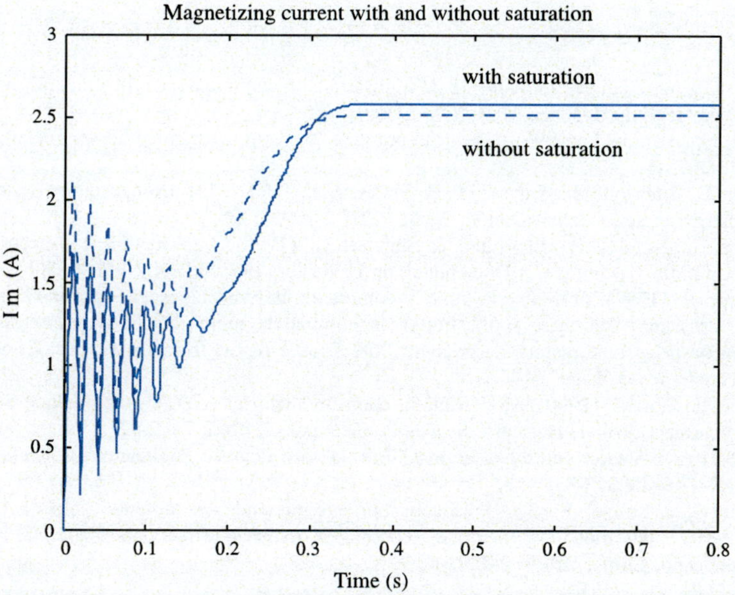

Fig. 2.16 Magnetizing current during the no-load starting of an induction machine, *solid* line: with magnetic saturation, *dashed* line: without magnetic saturation

the effective magnetizing current. β and s are chosen in order to make the experimental curve coincide with the theoretical one (Fig. 2.15). The equation of the magnetic characteristic is obtained by successive tests.

In order to demonstrate the importance of the magnetic saturation, Fig. 2.16 represents the magnetizing current obtained with and without the magnetic saturation (respectively in dashed and solid lines), and simulated when starting a no-load operation with an induction machine directly attached to the network. When the magnetic saturation is taken into consideration, the Fig. 2.16 shows that, in the beginning, the current peaks are less important, and that the magnetizing current is higher under steady conditions.

2.6 Conclusion

The physical and mathematical developments that are dealt with in this chapter have lead to determine a mathematical model reduced into a rotating frame of reference. The model parameters can be measured with different experimental procedures explained in [Caron and Hautier 1995]. The described modeling method can be used similarly for the modeling of three-phase induction motors [Sturtzer and Smigel 2000; Grenier et al. 2001]. The model.ing obtained in the Park reference frame will be used in the next chapter to elaborate on different speed control systems.

References

Caron, J.P., & Hautier, J.P. (1995). Modélisation et commande de la machine asynchrone, Editions technip, ISBN 2-7108-0683-5, Paris.

Rotella, F., & Borne, P. (1995). Théorie et pratique du calcul matriciel, éditions Technip, ISBN 2-7108-0675-4.

Grenier, D., Labrique, F., Buyse, H., & Matagne, E. (2001). Electromécanique—convertisseur d'énergie et actionneurs, Dunod, Paris, ISBN 2-10-005325-6.

Lesenne, J., Notelet, F., Labrique, & Séguier, G. (1995). Introduction à l'électrotechnique approfondie, Technique et Documentation Lavoisier, 1995, ISBN 2-85206-089-2.

Degobert, P. (1997). Formalisme pour la commande des machines électriques alimentées par convertisseurs statiques, application à la commande numérique d'un ensemble machine asynchrone—commutateur de courant, PhD Thesus report from Université des Sciences et Technologies de Lille.

Canudas de Wit, C. (2000). Modélisation, contrôle vectoriel et DTC—commande des moteurs asynchrones, Vol. 1. Hermes Sciences Publications, ISBN 2-7462-0111-9.

Vas, P. (1990). Vector control of ac machines, Oxford Science Publications, Clarendron Press, ISBN 0-19-859370-8.

Sturtzer, G., & Smigel, E. (2000). Modélisation et commande des moteurs triphasés, Commande vectorielle des moteurs synchrones, commande numérique par contrôleur DSP, collection Technosup, Ellipse, ISBN 2-7298-0076-X.

Chapter 3
Vector Control of Induction Machines

From the analysis carried out in Chap. 2, we can observe that the operation of the induction machine and its mathematical model is rather complicated. A certain number of specificities can be problematic, especially the following:

- electrical quantities are three-phased and alternating, thus leading to the use of phasors to represent voltage, current and stator and rotor flux;
- the coupling between the rotor and stator phases creates mutual inductances, which vary depending on the rotor position, thus generating torque;
- as the operation principle of this machine is induction, a part of the stator currents is used to magnetize it.

Furthermore, it should be added that it is not possible to measure rotor currents in squirrel-cage machines.

All these elements are naturally obstacles when designing a control system able to provide a drive system operating under variable speed. Indeed, many industrial processes use electric machines in elementary applications such as pumping, ventilation, convoying, machining, etc. In these applications, being able to regulate the speed means comfort of use and adaptation of mechanical energy to real needs, as well as a means to carry out technical–economic optimization during the industrial process. This chapter will deal exclusively with squirrel-cage machines.

Designing a system control device is all the more complicated as the system itself is complex. This is why, in the first part of this chapter, we will show that using C.O.G. formalism enables the systematization of the design approach and the standardization of a control system structure.

Most research work in the field of induction machine control has been dedicated to improving performances of the torque and flux variables dynamic control. First to appear was the scalar control based on simple static laws derived from a model of induction machine operating under permanent regime. Later on, starting from the differential equations resulting from a dynamic model of the machine, the so-called dynamic "vector" controls based on a study reference frame adjusted with a flux vector selected within the air gap became predominant. This strategy will be

B. Robyns et al., *Vector Control of Induction Machines*, Power Systems,
DOI: 10.1007/978-0-85729-901-7_3, © Springer-Verlag London 2012

explained in this chapter. It will be illustrated by the design of the control device of a simple actuator, the direct-current machine, and it will then be applied to the control of the induction machine.

In a second part, we will show how, thanks to a particular orientation of Park's reference frame, the torque expression of the induction machine becomes similar to the torque produced by a direct-current machine.

In a third part, the orientation is supposed to be carried out on the rotor flux, so as to cancel one of its components. The resulting simplified model is then presented and illustrated by a C.O.G. We will show that reversing this graph enables to determine the machine's direct vector control structure. The necessary control functions, compensations and estimators are also detailed. Controlling the torque developed by the machine requires controlling the flux. As the use of sensors that are able to measure these flux is not financially favorable, we will introduce different techniques that can be used to estimate these flux. Estimating and controlling electromagnetic torque are essential actions for system tractions, for instance locomotives. Estimating mechanical speed can be useful in an increasing number of applications, which seek to avoid the use of speed sensors. As an example, we will introduce the design of a rotor-flux reduced-order and full-order observer based on the formalism of state representation.

3.1 Formalism for the Design of Control Systems

3.1.1 Inverted Model Concepts

This part will introduce the foundations that will naturally allow structuring the control system. The design of a control, no matter the method or the form (continuous, sampled algorithm, fuzzy inferences, etc.), is an implicit study of an inverted model of the process to control. In other words, it is the expression of a will of causal inversion: 'as we know the effect produced by the cause, it is enough to create the right cause to obtain the desired effect.'

Thanks to C.O.G. formalism, the mathematical relations used to quantify the cause-to-effect physical phenomena have been represented in the shape of oval elementary processors. For a given elementary processor, the inversion of a cause-to-effect relation determines a control relation for the physical device thus modeled (Hautier and Faucher 1996). Control thus comes down to permuting the orientation of the variables in question and determining the process' inverted mathematical model. Reference quantities are marked by the index 'ref'. They refer to a desired trajectory for the effect which becomes a controlled quantity (or process output).

Adjustment quantities are marked by the index 'reg'. They are generated through the control relation and represent the necessary cause trajectory.

Depending on the nature and complexity of the mathematical model to invert, two types of relations can be inversed.

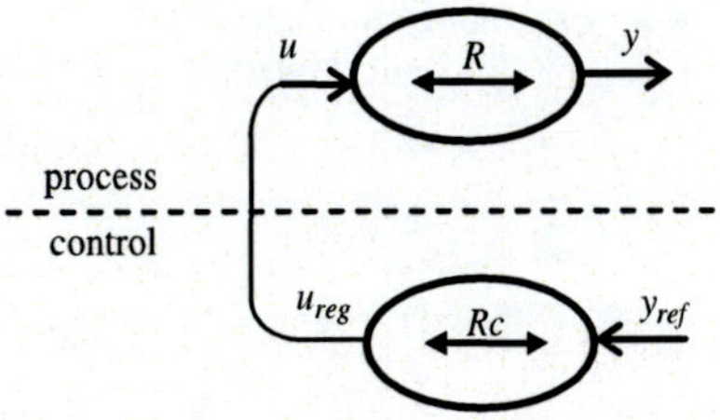

Fig. 3.1 Control by direct inversion of an instantaneous relation

3.1.2 Direct Inversion of an Instantaneous Relation

The expression 'instantaneous (or rigid) relation' refers to a time-independent, bijective (to a given value of cause u corresponds a unique value of effect y) mathematical relation instantly linking effect to cause:

$$R \rightarrow y = R(u) \tag{3.1}$$

The control relation aims at determining the cause to apply (u_{reg}) to impose a reference value (y_{ref}) to the desired effect (Fig. 3.1):

$$R_c \rightarrow u_{\text{reg}} = R_c(y_{\text{ref}}) \tag{3.2}$$

Given that the relation (R) is bijective, the control relation (Rc) is the inverted relation. If $u = u_{\text{reg}}$ and $Rc = R^{-1}$, then $y = y_{\text{ref}}$.

3.1.3 Indirect Inversion of a Causal Relation

Energy accumulation in the course of time makes the evolution of other influenced quantities dependent on time as well. This type of dynamic relation is called causal relation and is based on time-dependent differential relations:

$$\boxed{R \rightarrow y = R(u, t)} \tag{3.3}$$

Under these conditions, a causal relation cannot be bijective. The evolution of the influenced quantity results from a mathematical integration (of its differential equation). Forcing the evolution of this quantity requires the use of a derivation equation, which is, essentially, a physically impracticable operation, and subject to noise and numerical instability in practice. The reversal of a causal relation is then indirectly carried out by continuously taking into consideration the deviation $y_{\text{ref}} - y$ between the sensed value of the quantity y and the reference value y_{ref} (Hautier and Caron 1997). The role of the control relation is to reduce this deviation: this is the classic concept of a closed-loop control (Fig. 3.2):

$$\boxed{Rc \rightarrow u_{\text{reg}} = Rc(y_{\text{ref}} - y)} \tag{3.4}$$

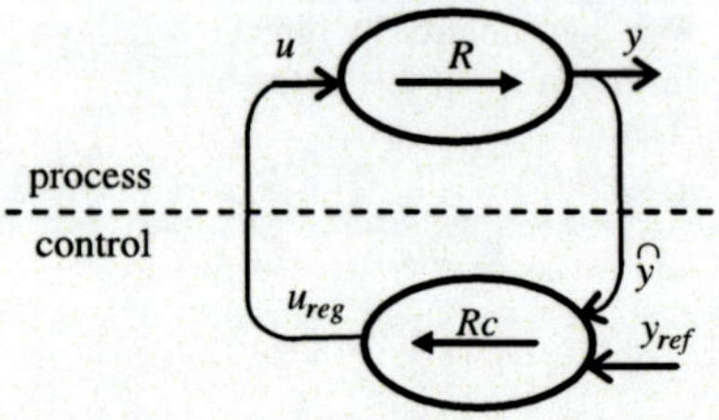

Fig. 3.2 Control by indirect inversion of a causal relation

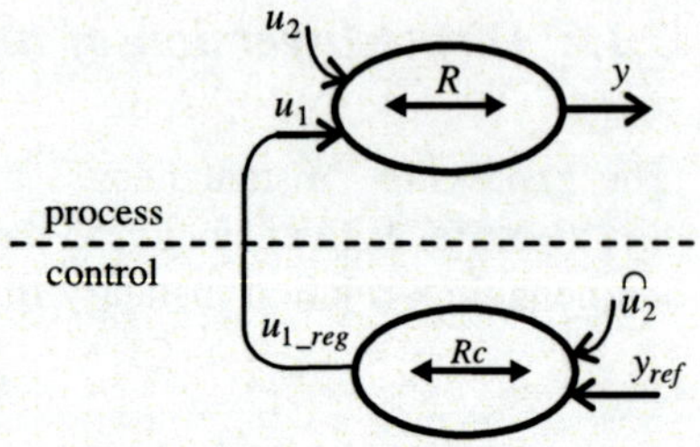

Fig. 3.3 Control by direct inversion of a multiple input relation

If $u = u_{\text{reg}}$ and $u_{\text{reg}} = C(y_{\text{ref}} - y)$, then $y = y_{\text{ref}}$, if and only if $C \to \infty$. Therefore *Rc* is a high-gain relation on the deviation. The symbol ∩ is used to indicate a measured variable. In this type of relations, the output quantity has to be either measured or estimated.

3.1.4 Indirect Inversion of a Relation with Various Inputs

Through a relation, an effect can depend on several causes. In that case, a single input (cause) is enough to control the output (produced effect). However, other inputs also have an influence on output and must be considered as disturbance inputs. The control relation must therefore take these inputs into account in order to compensate their action (Fig. 3.3).

Example 1 Considering the sum of two inputs

$$R \to y = u_1 + u_2 \tag{3.5}$$

If we choose to control the output y by acting on input u_1, then the control relation is expressed as follows:

$$Rc \to u_{1\_reg} = y_{ref} - \widehat{u}_2 \tag{3.6}$$

Example 2 Considering the product of two inputs

$$R \to y = u_1\, u_2 \tag{3.7}$$

Table 3.1 Causal organization of mathematical equations

Modeling relations	Control relations
$R1: u_L = u - u_R$	$R1c: u_{reg} = u_{Lreg} + \widehat{u}_R$
$R2: \phi = \int_{t_0}^{t_0+\Delta t} u_L \,\mathrm{d}t + \phi(t_0)$	$R2c: u_{Lreg} = C(\phi_{reg} - \tilde{\phi})$
$R3: i = \frac{1}{L}\phi$	$R3c\phi_{reg} = Li_{reg}$
$R_4: u_R = Ri$	

If we choose to control the output by acting on input u_1, then the control relation is expressed as follows:

$$Rc \rightarrow u_{1\_reg} = \frac{y_{ref}}{\widehat{u}_2} \tag{3.8}$$

This control relation requires all disturbance inputs to be measured.

3.1.5 Design of a Control Law

3.1.5.1 Maximum Control Structure

An electric machine is a system graphically represented by C.O.G. formalism as an assembling of two types of elementary processors. To design the system's control, each process C.O.G. processor is linked to a control system C.O.G. processor, which represents a control relation (Hautier and Faucher 1996). The reversal process can start from the main variable to control, which is then linked to a reference input of the control system. Starting from the quantity to control, the process's causal path is followed backwards and the processors' characteristic relations are inverted, until the control quantity is matched.

An instantaneous single-input single-output processor of the process C.O.G. corresponds to an instantaneous single-input single-output processor of the control system C.O.G. (Fig. 3.1).

A causal single-input single-output processor of the process C.O.G. corresponds to a causal single-input single-output processor of the control system C.O.G. (Fig. 3.2).

Thanks to this methodology of control system design, it is possible to keep the level of energy specific to each processor under control.

In the series R-L circuit example (Chap. 1, Fig. 1.10), the path between the control input (u) and the output to control (i) is defined by relations $R1$, $R2$ and $R3$ (Table 3.1, Fig. 3.4). The control system will therefore be based on three elementary processors $R1c$, $R2c$ and $R3c$. The equivalent bloc diagram representation is shown in Fig. 3.5. In a complex system, a control quantity can become a reference quantity if several control relations are cascading.

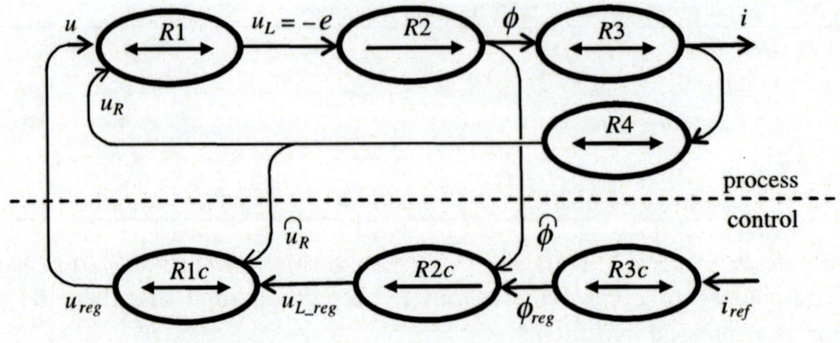

Fig. 3.4 C.O.G. of the model and of the maximum control structure

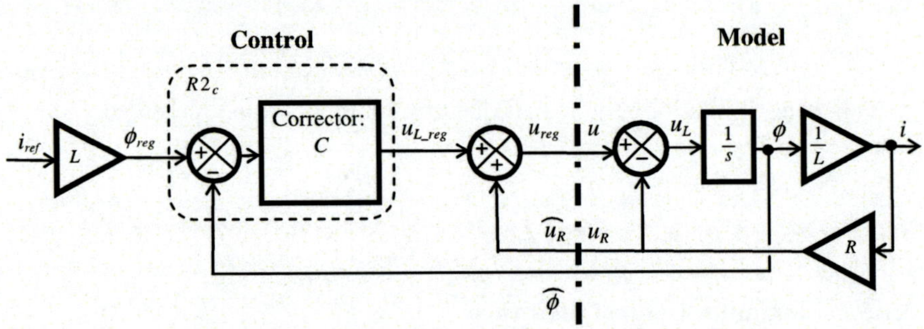

Fig. 3.5 Block diagram of the model and of the maximum control structure

If the C.O.G. processor of the process has several inputs and a causal output, it is essential to determine each input either as a control or as a disturbance input. In the given example, relation $R1$ includes an influenced input (u) and a disturbance input (u_R).

Measurable disturbance inputs are used as inputs for the control system's C.O.G. processor (relation $R1c$): this is called the compensation principle. This design approach results in the most elementary control structure, which implies that all quantities are measurable.

3.1.5.2 Control Structure Reduced to Measurable Quantities

If some physical quantities are not measurable, then the same approach can be used on the redesigned graph, showing only measurable quantities. In the given example, if the flux quantity is not measurable, then the process graph can be modified by grouping relations $R2$ and $R3$ within a single relation (Fig. 3.6):

$$\boxed{R2' \rightarrow i = \frac{1}{L}\int_{t_0}^{t_0+\Delta t} u_L \, dt + i(t_0)} \tag{3.9}$$

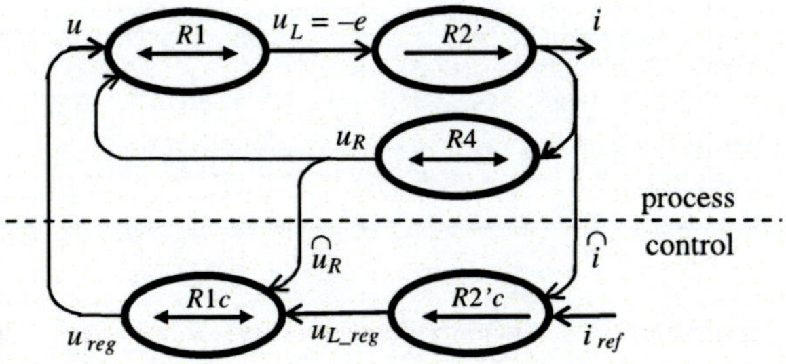

Fig. 3.6 C.O.G. of the model and of its control using measurable quantities

By inverting the path between single input (u) and desired output (i), the control system's structure only relies on two processors ($R1c$ unchanged and $R2'c$).

$$\boxed{R2'c \rightarrow u_{L\_\text{reg}} = C(i_{\text{ref}} - \overset{\frown}{i})} \tag{3.10}$$

The block diagram representation is shown in Fig. 3.7 with corresponding relations in Table 3.2. When disturbance inputs are not measurable, the structure of the control system has to be fundamentally modified. Three different structures must be distinguished: estimator-based control structure, observer-based control structure and controller-rejector-based control structure.

3.1.5.3 Control Structure Based on Estimators

In the given example, if u_R is a non-measurable disturbance, then it can be estimated using a relation equivalent to the relation resulting from the model and for which the input and output values are adapted:

$$\boxed{R4e \rightarrow \tilde{u}_R = R\overset{\frown}{i}} \tag{3.11}$$

The tilde ($\sim$) is used for estimated variables only. The other control relations remain unchanged (Fig. 3.8). It should be noted that the estimated relation is causal if the model's corresponding relation is. We will then speak about dynamic estimator.

3.1.5.4 Control Structure Based on observers

In numerous applications, there can be a lack of information about the origin of a non-measurable disturbance and its characterization through mathematical relations. When a non-measurable disturbance cannot be estimated, it can be observed by creating a closed-loop structure, which can make the estimated value converge toward the measured value.

Table 3.2 Causal organization of mathematical equations

Modeling relations	Control relations
$R1 : u_L = u - u_R$	$R1c : u_{\text{reg}} = u_{L\text{reg}} + \hat{u}_R$
$R2' : i = \frac{1}{L}\int_{t_0}^{t_0+\Delta t} u_L \, dt + i(t_0)$	$R2'c : u_{L\_\text{reg}} = C(i_{\text{ref}} - \hat{i})$
$R4 : u_R = Ri$	

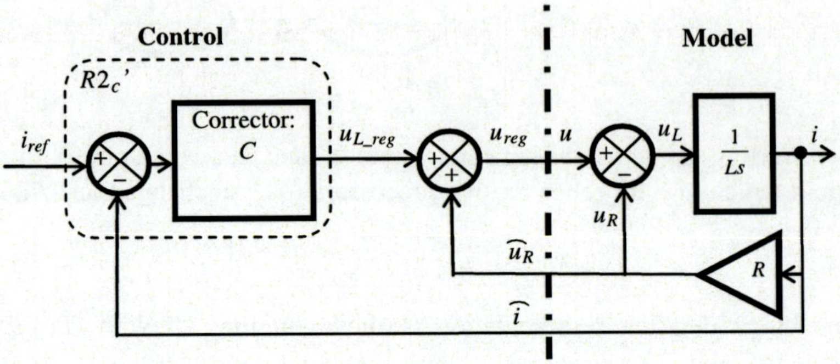

Fig. 3.7 Block diagram of the model and the control using measurable quantities

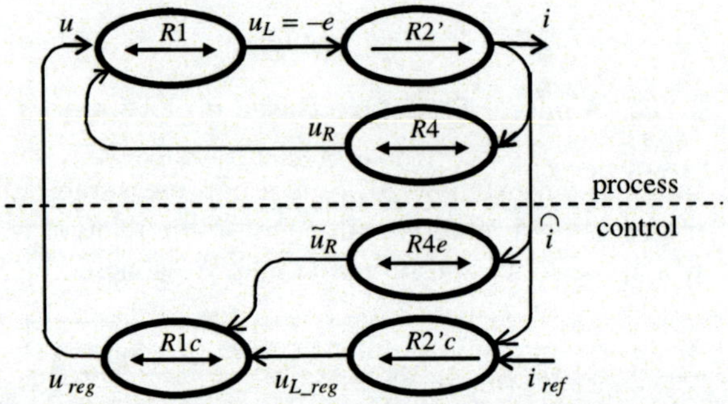

Fig. 3.8 C.O.G. of the small-scale model and its control based on static estimators

For instance, if a disturbance p appears at the inductance terminals, the estimated output $\tilde{i}$ is inevitably different from its measured value $\tilde{i}$ (Fig. 3.9). An observation of this disturbance can be obtained by reinjecting through a gain k the error $\varepsilon = \hat{i} - \tilde{i}$

$$\boxed{R5o \rightarrow \tilde{p} = k(\hat{i} - \tilde{i})} \tag{3.12}$$

$$\boxed{R2'e \rightarrow \tilde{i} = \frac{1}{L}\int_{t0}^{t0+\Delta t} \tilde{u}_L \, dt + \tilde{i}(t_0)} \tag{3.13}$$

$$\boxed{R4'e \rightarrow \tilde{u}_R = R\tilde{i}} \tag{3.14}$$

$$\boxed{R1e \rightarrow \tilde{u}_L = u_{\text{reg}} - \tilde{u}_R - \tilde{p}} \tag{3.15}$$

Relation $R1c$ is modified as follows: $R1c \rightarrow u_{\text{reg}} = u_{L\_\text{reg}} - \tilde{u}_R - \tilde{p}$.

The C.O.G. corresponding to this control structure is shown in Fig. 3.9. Voltage estimation $\tilde{u}_R$ resulting from current measurement ($R4e$) is, with a standard quality sensor, more precise than voltage estimation resulting from the estimator ($R4'e$) Table 3.3.

3.1.5.5 Control Structure Based on Controller Rejectors

Using a control structure based on observers can make the control system in equations considerably unwieldy. An alternative consists in reducing the effects of the non-measurable, unobservable disturbance inputs on the output, by properly designing the controller used in the closed loop (equivalent to causal processor): this principle is known as disturbance rejection through loop gain increase, commonly called regulator.

In the given example, if u_R is a non-measurable and unobservable disturbance, relation $R1c$ becomes unnecessary and a unique relation for the control appears (Fig. 3.10):

$$\boxed{Rc \rightarrow u_{\text{reg}} = C(i_{\text{ref}} - \hat{i})} \tag{3.16}$$

This relation will reduce the influence of the non-measured disturbance, as well as adjust the output to the reference. A global equation of the model, whose resolution determines the effect, can then be used:

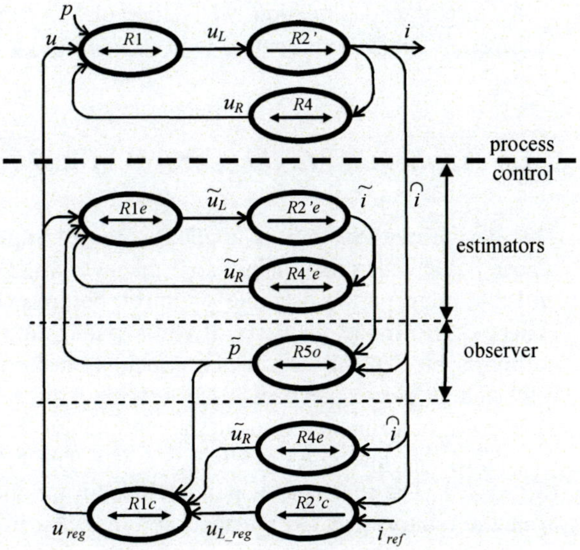

Fig. 3.9 C.O.G. of the small-scale model and its control based on observers

Table 3.3 Causal organization of mathematical equations

Modeling relations	Estimating relations	Control relations
$R1 : u_L = u - u_R - p$	$R1e : \tilde{u}_L = u_{\text{reg}} - \tilde{u}_R - \tilde{p}$	$R1c : u_{\text{reg}} = u_{L\text{reg}} + \tilde{u}_R + \tilde{p}$
$R2' : i = \frac{1}{L}\int_{t_0}^{t_0+\Delta t} u_L \, \mathrm{d}t + i(t_0)$	$R2'e : \tilde{i} = \frac{1}{L}\int_{t_0}^{t_0+\Delta t} \tilde{u}_L \, \mathrm{d}t + \tilde{i}(t_0)$	$R2'c : u_{L\_\text{reg}} = C(i_{\text{ref}} - \hat{i})$
$R4 : u_R = Ri$	$R4e : \tilde{u}_R = R\tilde{i}$	$R4e : \tilde{u}_R = R\hat{i}$
	observing realation	
	$R5o : \tilde{p} = k\left(i - \tilde{i}\right)$	

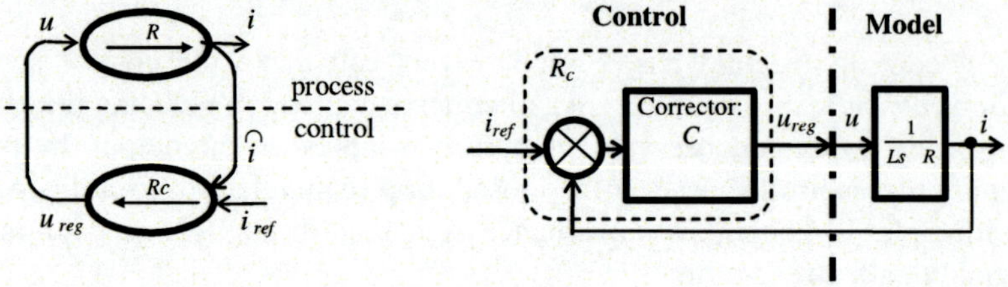

Fig. 3.10 C.O.G. and block diagram of the small-scale model and its control based on controller rejectors

$$\boxed{R \rightarrow L\frac{\mathrm{d}i}{\mathrm{d}t} + Ri = u} \tag{3.17}$$

This control structure is the most widely used in industry. It must be noted that for some applications, it is not possible to obtain performances similar to those of previous structures. Besides, the complexity of the controller is much more important.

An instantaneous relation, whose parameters can vary or be unknown, can also be inverted thanks to a control loop (thus through this indirect method).

3.1.6 Application to the Control of a Direct-Current Machine

This design methodology is applied to the example of a direct-current machine, whose inductor is made thanks to a magnet, which generates a constant flux. We will first focus on the machine's control, before extending it to the vector control concept. The model consists of three parts: an electrical part fit for armature coiling, a mechanical part which depends on the mechanical load driven by the rotor and an electromechanical conversion (Louis 2004) (Fig. 3.11).

The use of formalism leads to the mathematical relations of Table 3.4 and to the C.O.G. of this machine model (Fig. 3.12). The block diagram is easily deduced (Fig. 3.13). The mechanical part includes a mobile armature (rotor) and a fixed armature (stator) and has already been modeled in Chap. 1 (Sect. 1.4.2.4).

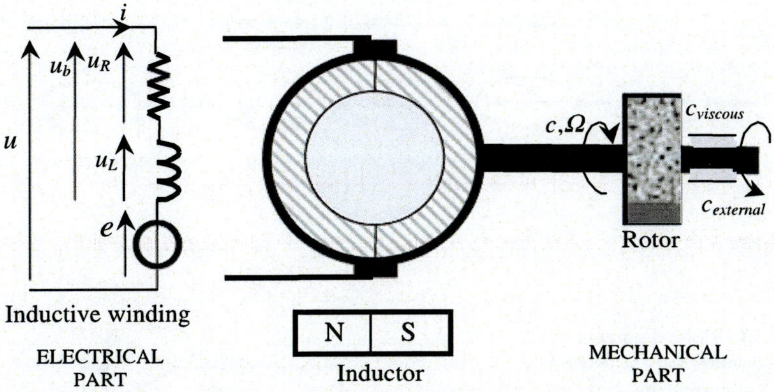

Fig. 3.11 Functional breaking down of a direct-current machine

Table 3.4 Causal organization of direct current machine equations

Inductive circuit	Electromechanical conversion
$R1 : u_b = u - e$	$R6 : T = k\,i$
$R2 : u_L = u_b - u_R$	$R7 : e = k\,\Omega$
$R3 : \phi = \int_{t_0}^{t_0+\Delta t} u_L \mathrm{d}t + \phi(t_0)$	
$R4 : i = \frac{1}{L}\phi$	
$R5 : u_R = Ri$	

The only quantities considered as physically accessible are current and speed, respectively noted as $\widehat{i}$ and $\widehat{\Omega}$. Relations $R1$, $R2$, $R3$, $R4$ and $R5$ lead to the following differential equation:

$$\boxed{R1' \to L\frac{\mathrm{d}i(t)}{\mathrm{d}t} + Ri(t) = u(t) - e(t)} \tag{3.18}$$

Likewise, all the equations modeling the mechanical part and the load (Eq. 1.76) can be summed up by the following differential equation:

$$\boxed{R7' \to J_r\frac{\mathrm{d}\Omega(t)}{\mathrm{d}\tau} + f_r\,\Omega(t) = T(t) - T_{\mathrm{ch}}(t)} \tag{3.19}$$

The C.O.G. of the model based on measurable quantities is shown in Fig. 3.14. Input quantities for the process are armature voltage u load torque c_{ch} and torque coefficient k, which could be linked to a current if the inductor is coiled.

The lower part of Fig. 3.14 shows the C.O.G. model of the speed control of a direct-current machine. The structure can be obtained by inverting causalities.

Three variables have an effect on torque: armature current i, torque coefficient k (which represents, give or take a coefficient, the flux generated by the magnets)

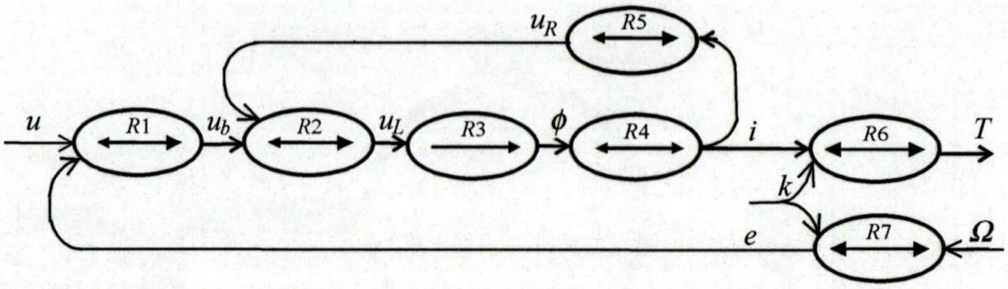

Fig. 3.12 C.O.G. of the electromechanical part model of the direct-current machine

Fig. 3.13 Block diagrams of the electromechanical part of the direct-current machine

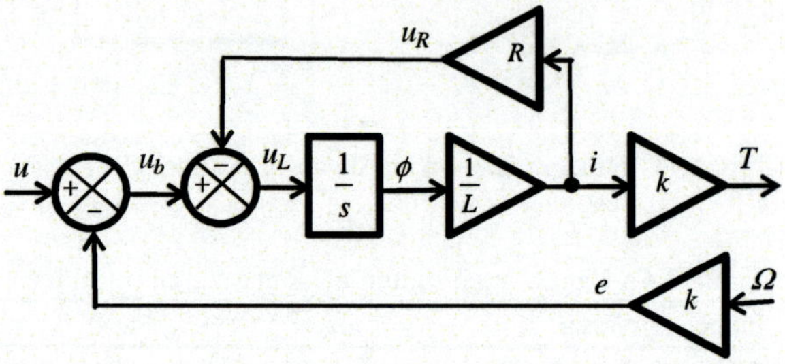

and brush stagger angle in keeping with the armature's neutral line. Indeed, torque actually results from the vector product of inductor flux and armature current. This product is optimal when the armature conductors are perpendicular to the field lines (Degobert 1997; Hautier and Caron 1997).

Control relations are then almost invariably deducted from process control relations.

We distinguish:

- armature current control with electromotive force compensation:

$$R1'c \rightarrow u_{reg} = C_i\,(i_{ref} - i) + \tilde{e} \tag{3.20}$$

- electromotive force estimate, which is a simple copy of the relation used in the model:

$$R6e \rightarrow \tilde{e} = k\widehat{\Omega} \tag{3.21}$$

- calculation of the current reference:

$$R5c \rightarrow i_{reg} = \frac{1}{k} T_{reg} \tag{3.22}$$

- speed control:

$$R7' \rightarrow T_{reg} = C_\Omega\left(\Omega_{ref} - \widehat{\Omega}\right) \tag{3.23}$$

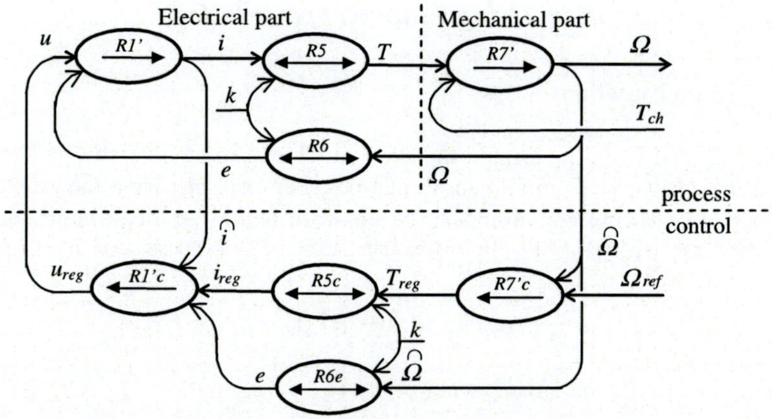

Fig. 3.14 C.O.G. of the model and of the control of a direct-current machine

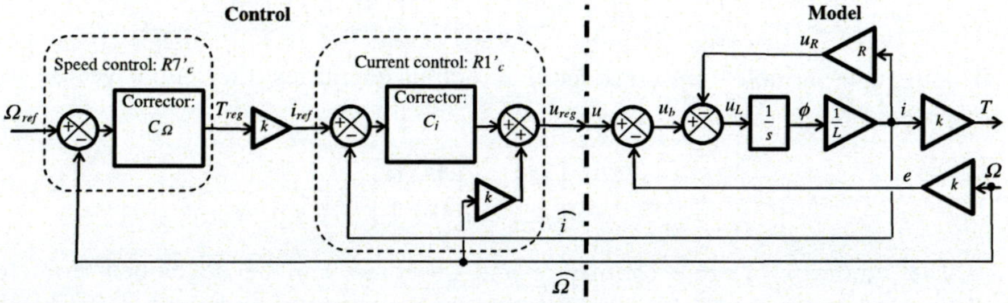

Fig. 3.15 Block diagram of the direct-current machine control

For a constant flux direct-current machine (regulated field current or permanent magnets), torque control is equivalent to the armature current control.

The control device bloc diagram is shown in Fig. 3.15.

E.m.f. compensation is useless under steady conditions, especially if the controller is a proportional-integral controller. The same applies to transient conditions in which the speed has little variations: for instance with variable load speed regulation. Nevertheless, under rapidly shifting speed conditions (position tracking), this compensation can help the controller, whose only role is to determine voltage drop at the inductance (and resistance) terminals if the ohmic compensation takes place (or not) (Louis et al. 1999). It could thus be demonstrated that, to obtain given performances in tracking, the controller dynamics can be reduced in the presence of the e.m.f. compensation, which would make the control loop less noise-sensitive.

3.1.7 Generalization Using State Representation

3.1.7.1 State Equations

Using a Causal Ordering Graph, the state variables can be obtained by extracting the output derivative from all causal relations. For example, from the graph of the model, the following equations can be obtained using the measurable quantities (Fig. 3.14):

$$\frac{\mathrm{d}\Omega(t)}{\mathrm{d}t} = -\frac{f}{J_r}\Omega(t) + \frac{1}{J_r}T(t) - \frac{1}{J_r}T_{\mathrm{ch}}(t) \tag{3.24}$$

$$\frac{\mathrm{d}i(t)}{\mathrm{d}t} = -\frac{R}{L}i(t) + \frac{1}{L}u(t) - \frac{1}{L}e(t) \tag{3.25}$$

If we use vector notations for the state quantities, we obtain the following:

$$\begin{bmatrix} \dot{\Omega} \\ \dot{i} \end{bmatrix} = \begin{bmatrix} -\frac{f}{J_r} & \frac{k}{J_r} \\ -\frac{k}{L} & -\frac{R}{L} \end{bmatrix} \begin{bmatrix} \Omega \\ i \end{bmatrix} + \begin{bmatrix} -\frac{1}{J_r} & 0 \\ 0 & \frac{1}{L} \end{bmatrix} \begin{bmatrix} T_{\mathrm{ch}} \\ u \end{bmatrix} \tag{3.26}$$

If both state variables are considered as output quantities, the output vector corresponds to:

$$[Y] = \begin{bmatrix} \Omega \\ i \end{bmatrix} = \begin{bmatrix} 1 & 0 \\ 0 & 1 \end{bmatrix} [X] \tag{3.27}$$

The synthesis of state estimators and observers can be developed from the differential equations of the model, written in vector form and synthesized by the following state equations (Borne et al. 1990):

$$[\dot{X}] = [A]\,[X] + [B]\,[U] \tag{3.28}$$

$$[Y] = [C]\,[X] + [D]\,[U] \tag{3.29}$$

$[X]$ stands for the state-variables vector, $[U]$ stands for the input vector and $[Y]$ stands for the output vector. Matrixes $[A]$, $[B]$, $[C]$ and $[D]$ depend on the model of the system. These matrixes are time-independent in a linear system.

3.1.7.2 State Estimators

The principle of estimators consists in using the model equations to calculate an estimate of a non-measurable quantity (Sect. 3.1.5.3) (Luenberger 1966). If we take the example of output quantities, a static estimator of these outputs will be obtained by using the mathematical equations deriving from the model and matching inputs and outputs to the quantities deriving from the control:

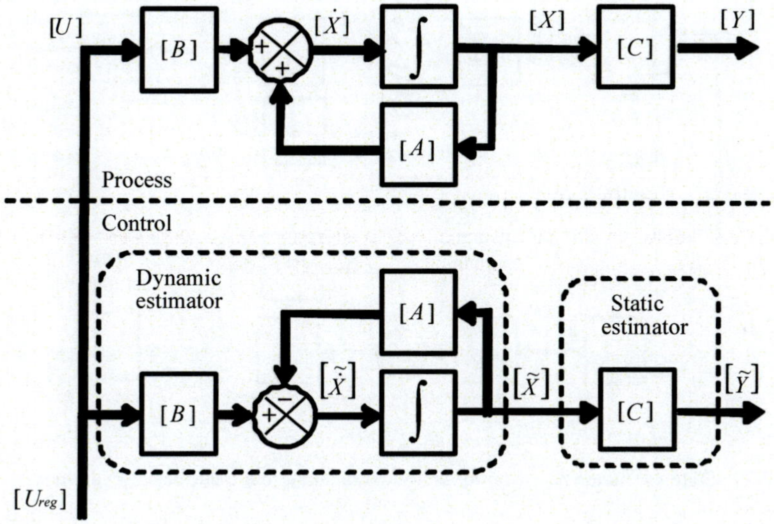

Fig. 3.16 Determining static and dynamic estimators

$$\left[\tilde{Y}\right] = [C]\left[\tilde{X}\right] \tag{3.30}$$

If we take the example of state quantities, a dynamic estimator of these states can also be obtained by using the dynamic equations deriving from the model and matching inputs and outputs to the quantities deriving from the control:

$$\left[\dot{\tilde{X}}\right] = [A]\left[\tilde{X}\right] + [B]\left[U_{reg}\right] \tag{3.31}$$

Those two estimators are called open-loop estimators and can be controlled by the control inputs $[U_{\text{reg}}]$ deriving from the control device (Fig. 3.16).

In general, the input vector $[U]$ includes control inputs (written $[U_i]$) and disturbance inputs (written $[Up]$):

$$[U] = \begin{bmatrix} [Ui] \\ [Up] \end{bmatrix} \tag{3.32}$$

If all disturbance inputs can be measured, then the structure can easily be modified to take into account the incidence of these disturbance inputs on the observation and estimation (Fig. 3.17):

$$\left[\tilde{U}\right] = \begin{bmatrix} \left[U_{reg}\right] \\ \left[\widehat{U}p\right] \end{bmatrix} \tag{3.33}$$

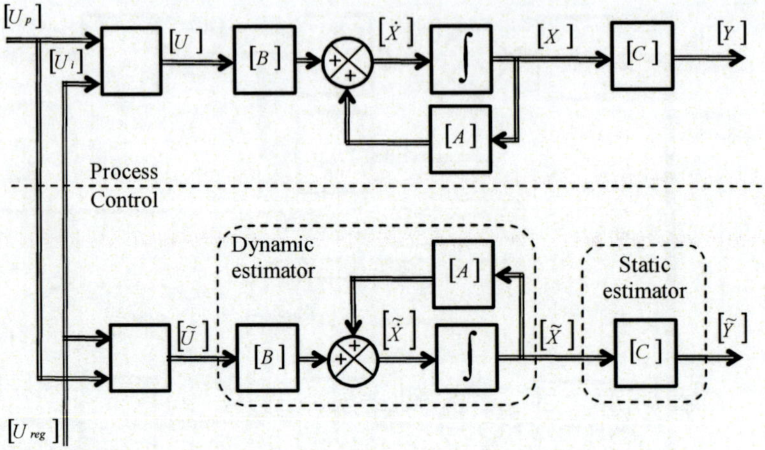

Fig. 3.17 State estimator based on measured adjustment and disturbance quantities

3.1.7.3 State Observers

Observers are used when part of the model is not well-known (Sect. 3.1.5.4).

In case some of the disturbance inputs are not measurable, an error obviously remains between the output quantities and their estimates.

$$[\varepsilon] = \left[\widehat{Y}\right] - \left[\tilde{Y}\right] = \left[\widehat{Y}\right] - [C]\left[\tilde{X}\right] \tag{3.34}$$

This gap can therefore be used to make weighted corrections to the estimated value of the state vector derivative (Fig. 3.18). The obtained structure is called observer. It corresponds to a closed-loop estimator. If we add Eqs. 3.31–3.34, we obtain:

$$\left[\dot{\tilde{X}}\right] = [A]\left[\tilde{X}\right] + [B]\left[U_{reg}\right] + [K]\left[\left[\widehat{Y}\right] - [C]\left[\tilde{X}\right]\right] \tag{3.35}$$

Gain matrix $[K]$ is determined in order to adjust the observer dynamic, while determining the proper values of the resulting state matrix $[Ar] = [A] - [K]\,[C]$. This matrix sets the dynamics of the observation error $[\varepsilon]$. The fact that matrix $[K]$ highly influences the observer‘s sensitivity to uncertainties on electric parameters will be studied in the next chapters. Choosing the matrix $[K]$ is therefore often difficult, since the two criteria—dynamics and sensitivity—often lead to contradictory gain values.

3.2 Flux-Orientation Control Strategies

Like any other electromechanical converter, the induction machine is first and foremost a torque generator. We already learned that the electromagnetic torque could be expressed in different ways. However, all these expressions rely on a

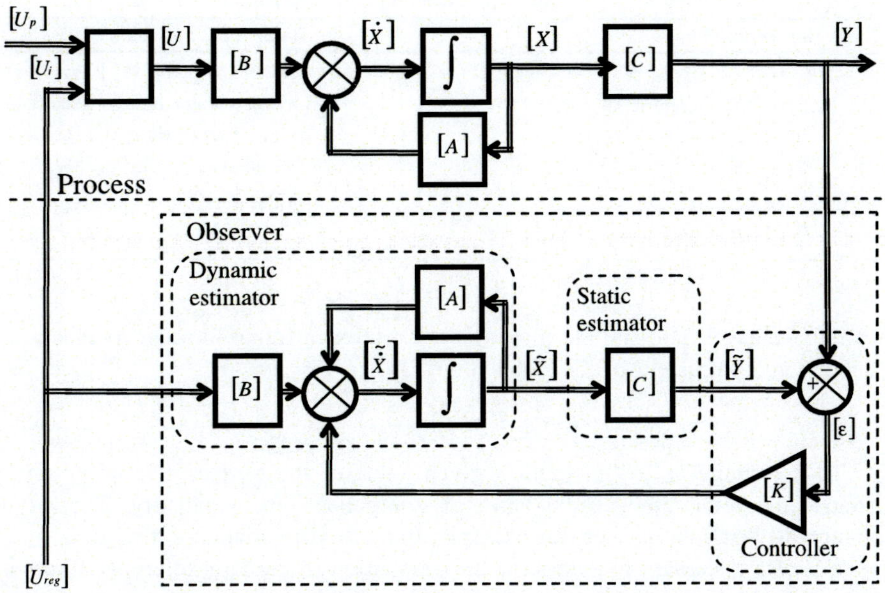

Fig. 3.18 Closed-loop state observer

common mathematical formula: the addition of two products (Sect. 2.4.2.4). Each product corresponds to a torque from a fictitious machine d and a machine q. The uniqueness of this relation leads to an infinity of solutions so as to divide the value of reference torque T_{ref} between the two virtual machines. However, this mathematical form can be significantly amplified by canceling one of the two terms (Table 3.5).

The vector control strategy therefore consists in orienting a Park reference frame (d, q) so as to cancel one of the flux components (direct or, more often, quadrature) in order to simplify the torque mathematical expression. This simplified expression is then used to determine torque control.

If the flux control is carried out on its two components in the Park reference frame, we talk about vector control with oriented flux. There are various equivalent torque expressions, which lead to various control algorithms aiming at controlling torque to a reference. Therefore, from torque expression in the rotor flux referential (Eq. 2.106), we can choose to locate the rotating Park reference frame so as to cancel the quadrature component of the rotor flux ($\phi_{rq} = 0$, $\phi_{rd} = \phi_r$). We then obtain a positive and simplified expression:

$$T = p \frac{M}{L_r} \left(i_{sq}\, \phi_{rd} \right) = T_{rq} \tag{3.36}$$

This technique is, by far, the most used and consists in aligning the d-axis with the rotor flux. From the torque expression in the stator flux frame (Eq. 2.107), we

Table 3.5 Classifying torque expressions depending on the flux used

Rotor flux expression	Stator flux expression
$T = p\frac{M}{L_r}\left(i_{sq}\phi_{rd} - i_{sd}\,\phi_{rq}\right)$	$T = p\left(i_{sq}\phi_{sd} - i_{sd}\phi_{sq}\right)$
Oriented rotor flux command	Oriented stator flux command
on the direct axis $\phi_{rq} = 0$	on the direct axis $\phi_{sq} = 0$
$T = p\frac{M}{L_r}\left(i_{sq}\,\phi_{rd}\right)$	$T = pi_{sq}\phi_{sd}$
on the quadrature axis $\phi_{rd} = 0$	on the quadrature axis $\phi_{sd} = 0$
$T = p\frac{M}{L_r}\left(i_{sq}\,\phi_{rd}\right)$	$T = -pi_{sd}\phi_{sq}$

can also choose to locate the rotating Park reference frame in order to cancel the quadrature component of the stator flux:

$$T = p(i_{sq}\phi_{sd}) \tag{3.37}$$

To determine at any moment a given value of the torque, it is necessary to control the remaining current and flux components independently. The torque value will bring an answer, for example, to a speed regulation constraint.

It is also necessary to determine the orientation of the Park reference frame by calculating the real-time values of angles θ_s and θ_r in order to set an orientation of the frame of reference that verifies the cancelation hypothesis of one of the two flux components. This is called the *feedback* principle.

A priori control consists in understanding the machine control from a simplified model, for which one of the two fictive torques is null.

3.3 Vector Control with Oriented Rotor Flux

3.3.1 A Priori Modeling on the Flux Orientation

3.3.1.1 Modeling and C.O.G. of the Model

Starting from the mathematical model using rotor flux (Sect. 2.4.4) and if we gather the dynamic equations determining the various flux (group of three equations pictured in gray in Fig. 2.10), we obtain relations $R1'_{sd}$, $R1'_{rd}$, $R1'_{rq}$ et $R1'_{sq}$ (Table 3.6).

The quadrature component of the rotor flux is supposed to be null; this hypothesis is therefore added to the machine equations pictured in Table 2.2 (Sect. 2.4.3.2). Since the rotor circuit of the machine studied is short-circuited, rotor voltages (v_{rd}, v_{rq}) are canceled in equations $R1_{rd}$ and $R1_{rq}$. If the quadrature component of the rotor flux is constantly null, then integral equation ($R2_{rq}$) implies that:

$$e_{\phi rq} = 0 \tag{3.38}$$

Table 3.6 Causal organization of Park reference frame with $\phi_{rq} = 0$

Electromagnetical part:	
$R1'_{sd}: \frac{d\phi_{sd}}{d_t} = v_{sd} - R_s\, i_{sd} - e_{sd}$	$R1'_{rd}: \frac{d\phi_{rd}}{d_t} = -R_r\, i_{rd}$
$R1'_{sq}: \frac{d\phi_{sq}}{dt} = v_{sq} - R_s i_{sq} - e_{sq}$	$R1'_{rq}: \frac{d\phi_{rq}}{d_t} = -R_r\, i_{rq} - e_{rq} = 0$
$R3_{sd}: i_{sd} = \frac{1}{L_s}(\phi_{sd} - Mi_{rd})$	$R3_{rd}: i_{rd} = \frac{1}{L_R}(\phi_{rd} - Mi_{sd})$
$R3'_{sq}: i_{sq} = \frac{1}{\sigma . L_s}\phi_{sq}$	$R3'_{rq}: i_{rq} = -\frac{1}{L_r} Mi_{sq}$
Electromechanical coupling:	
$R5_{sd}: T_{rd} = 0\ R5_{sq}: T_{rq} = p\frac{M}{L_r}\, i_{sq}\, \phi_{rd}$	
$R6_{sd}: e_{sd} = -\phi_{sq}\, \omega_s$	$R6_{rd}: e_{rd} = 0$
$R6: e_{sq} = \phi_{sd}\, \omega_s$	$R6_{rq}: e_{rq} = \phi_{rd}\, \omega_r$
$R7_s: c = c_{rq}$	
Referential speed:	$R11: \theta_r = \int_{t_0}^{t_0+\Delta t} \omega_r\, dt + \theta_r(t_0)$
$R8_s: \omega_s = p\Omega + \omega_r$	
$R10_s: \theta_s = \theta_r + \theta$	

Array equation ($R1_{rq}$) can be simplified into ($R1'_{rq}$). Also, the expression of the quadrature component of rotor current ($R3_{rq}$) can be simplified into ($R3'_{rq}$). If we add this relation into the expression of the quadrature component of stator current ($R3_{sq}$), we obtain:

$$R3'_{sq} \rightarrow i_{sq} = \frac{1}{L_s}\left(\phi_{sq} + \frac{M^2}{L_r} i_{sq}\right) = \frac{1}{\sigma L_s}\phi_{sq} \tag{3.39}$$

Concerning the relations peculiar to axis d, the cancelation of the quadrature component of the flux leads to:

$$R6_{rd} \rightarrow e_{rd} = 0 \tag{3.40}$$

$$R1'_{rd} \rightarrow e_{\phi rd} = -v_{\rightarrow} = -R_r i_{rd} \tag{3.41}$$

All those equations are summed up in Table 3.6. The C.O.G. of the induction machine model (Fig. 2.12) can be simplified (Fig. 3.19).

From that graph, we can underline the fact that the direct component of rotor flux (ϕ_{rd}) can be controlled from the direct component of stator current (i_{sd}). The quadrature component of the stator current can therefore be used to adjust the torque. These two currents can themselves be adjusted from the stator voltages.

3.3.1.2 Determining the Magnetizing Current

The magnetizing current is defined by:

$$i_{\phi r} = \frac{\phi_{rd}}{M} = \frac{\phi_r}{M} \tag{3.42}$$

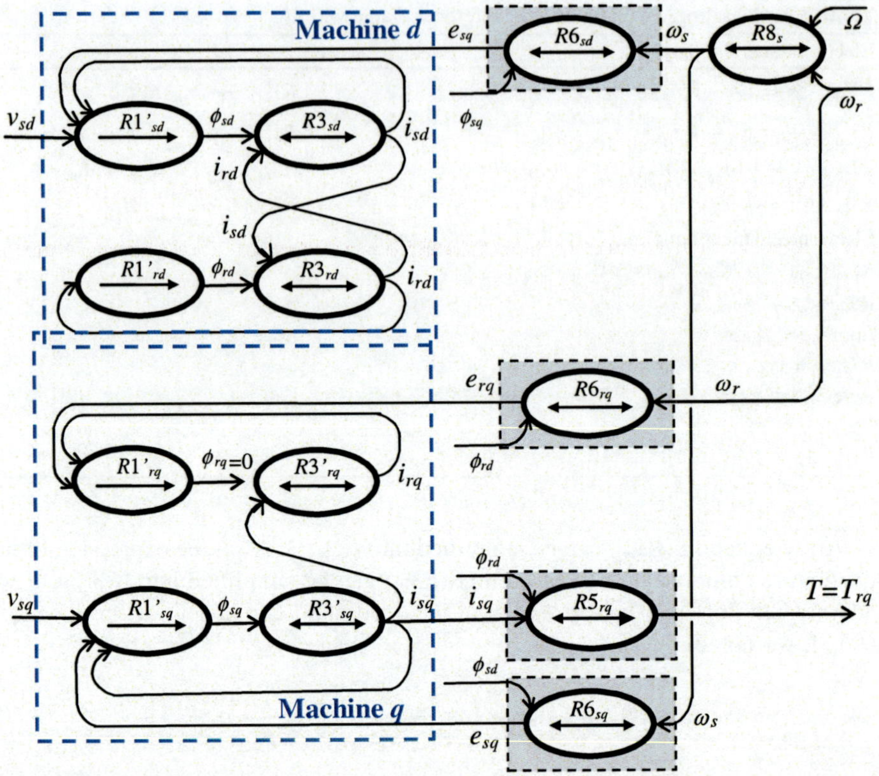

Fig. 3.19 Underlying the simplifications with $\phi_{\text{rq}} = 0$ (the relations are reported in Table 3.6)

The interest of this variable change is to obtain a model with only measurable parameters. Indeed, with an ordinary squirrel-cage machine, the values of cyclical inductance and rotor resistance (L_r, R_r) and cyclical mutual inductance (M) are not accessible and hardly ever given by the makers. Despite that, there are methods to measure the four classical parameters L_s, R_s, σ and τ_r.

The rotor flux (that equals ϕ_{rd} since $\phi_{\text{rq}} = 0$) depends only on current i_{sd} through relations $R1'_{\text{rd}}$ and $R3_{\text{rd}}$ (Fig. 3.19). Applying the Laplace transform (Barre et al. 1995) gives a clearer expression and shows the current determination:

$$\boxed{R16 \rightarrow i\phi r(s) = \frac{1}{1+\tau_r s}\, i_{\text{sd}}(s)} \tag{3.43}$$

Variable (s) represents the Laplace variable and τ_r is the time constant of the rotor current, with $\tau_r = L_r/R_r$.

Determining the stator current:

The direct component of the stator current only depends on the direct component of the voltage through relations $R3_{\text{sd}}$ and $R1'_{\text{sd}}$ (Fig. 3.19). Using the flux expression resulting from relation $R3_{\text{sd}}$, differential equation $R1'_{\text{sd}}$ becomes:

$$L_s \frac{\mathrm{d}i_{sd}}{\mathrm{d}t} + R_s i_{sd} = v_{sd} - e_{sd} - M \frac{\mathrm{d}i_{rd}}{\mathrm{d}t} \tag{3.44}$$

If we replace the derivative from relation $R3_{rd}$, in this equation, we deduce the following:

$$L_s\sigma \frac{\mathrm{d}i_{sd}}{\mathrm{d}t} + R_s i_{sd} = v_{sd} - e_{sd} - \frac{M}{L_r}\frac{\mathrm{d}\phi_{rd}}{\mathrm{d}t} \tag{3.45}$$

If the rotor flux is supposed to be controlled a priori, it is constantly equal to its reference and its derivative is null. Under steady conditions, the equation has to be changed into the following:

$$\boxed{R17 \rightarrow L_s\sigma \frac{\mathrm{d}i_{sd}}{\mathrm{d}t} + R_s i_{sd} = u_{sd}} \tag{3.46}$$

with:

$$\boxed{R18 \rightarrow u_{sd}(s) = v_{sd}(s) - e_{sd}(s)} \tag{3.47}$$

The direct component of the stator e.m.f. depends on the quadrature component of the non measurable stator flux ($R6_{sd}$). By adding relation $R3'_{sq}$ to $R6_{sd}$, we obtain an expression of this e.m.f. depending on the quadrature component of the stator current, which is a measurable quantity (Fig. 3.20):

$$\boxed{R19 \rightarrow e_{sd} = -\sigma L_s \omega_s i_{sq}} \tag{3.48}$$

If we apply the Laplace transform onto the direct component of stator current (Eq. 3.46) and replace it in Eq. 3.43, we obtain a global relation that is useful to determine the magnetizing current:

$$\boxed{R20 \rightarrow i_{\phi r}(s) = \frac{1}{R_s}\frac{1}{\sigma\tau_s\tau_r s^2 + (\tau_s + \tau_r)s + 1} u_{sd}(s)} \tag{3.49}$$

τ_s is the following time constant: $\tau_s = L_s/R_s$. This relation can therefore be a substitute to relations $R16$ and $R17$ in Fig. 3.20.

3.3.1.3 Determining the Torque

Using the expression of the magnetic current (Eq. 3.42), the torque expression (Eq. 3.36) becomes:

$$\boxed{R12 \rightarrow T = p\frac{M^2}{L_r} i_{sq}\, i_{\phi r} = p(1-\sigma) L_s i_{sq} i_{\phi r}} \tag{3.50}$$

The quadrature component of the stator current results from a loop that implicates relations $R3'_{sq}$ and $R1'_{sq}$ (Fig. 3.19). This loop can be reduced to two relations

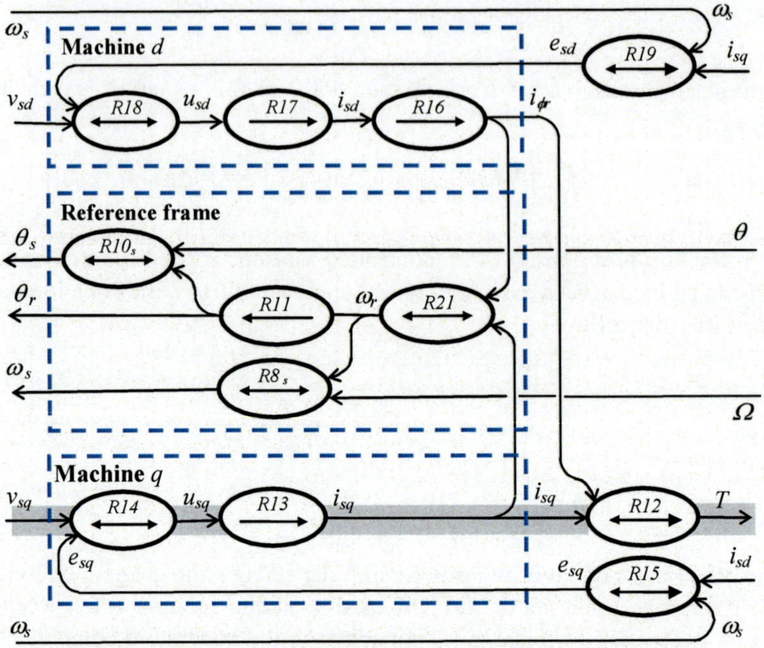

Fig. 3.20 C.O.G. of the simplified model with orientation of the rotor flux

($R13$ and $R14$) of which the control input (influenced cause) is the quadrature component of the stator voltage (Fig. 3.20):

$$\boxed{R13 \rightarrow i_{\mathrm{sq}}(s) = \frac{1}{R_s}\frac{1}{1+\sigma\tau_s s}u_{\mathrm{sq}}(s)} \tag{3.51}$$

with

$$\boxed{R14 \rightarrow u_{\mathrm{sq}}(s) = v_{\mathrm{sq}}(s) - e_{\mathrm{sq}}(s)} \tag{3.52}$$

$\sigma\tau_s$ is the time constant of the stator current. These relations are equal to an induction coil load σL_s with a resistance R_s and a voltage u_{sq}.

Calculating the quadrature component of the e.m.f.:

The expression of the quadrature component of the e.m.f. (relation $R6_{\mathrm{sq}}$, Table 3.4) uses the direct component of the stator flux, which, for usual machines, cannot be measured. However, an equivalent expression using the direct component of the stator flux can be determined. Relation ($R3_{\mathrm{sd}}$) can therefore be rewritten under the following form:

$$\phi_{\mathrm{sd}} = L_s i_{\mathrm{sd}} + M i_{\mathrm{rd}} \tag{3.53}$$

If we use relation ($R3_{\mathrm{rd}}$) to replace the direct component of the rotor current, we obtain the following:

$$\phi_{sd} = L_s i_{sd} + \frac{M}{L_r}\phi_{rd} - \frac{M^2}{L_r} i_{sd} = L_s \sigma i_{sd} + \frac{M^2}{L_r} i_\phi \tag{3.54}$$

The magnetizing current is not a measurable quantity. If we use its expression (Eq. 3.43), we get a stator flux expression that depends only on the stator current:

$$\phi_{sd} = L_s \frac{1 + \sigma\tau_r s}{1 + \tau_r s} i_{sd} \tag{3.55}$$

A second expression of the e.m.f. can therefore be used:

$$\boxed{R15 \rightarrow e_{sq}(s) = L_s \frac{1 + \sigma\,\tau_r s}{1 + \tau_r\, s}.(i_{sd}(\tau) \cdot \omega_s(\tau))} \tag{3.56}$$

3.3.1.4 Incidence of the Frame Orientation

The various hypotheses on the reference frame orientation and on the cancelation of the quadrature component of the rotor flux resulting from it lead to particular relations between quantities. The causal orientation of relation $R6_{rq}$ must be modified, because, in these conditions, the rotor pulsation depends both on voltage and on the quadrature component of the rotor flux (Table 3.4):

$$\omega_r = e_{rq} \frac{1}{M i_{\phi r}} \tag{3.57}$$

Thus, pulsation ω_r and current i_{sq} become linked by an instantaneous relation through relations $R1'_{rq}$ and $R3_{rq}$ (Table 3.4):

$$\boxed{R21 \rightarrow \omega_r = \frac{R_r}{L_r} \frac{1}{i_{\phi r}} i_{sq}} \tag{3.58}$$

The rotor angle is obtained when integrating rotor pulsation ($R11$). The stator pulsation (which is the speed of the frame of reference linked to the stator, Fig. 2.11) then depends on the rotor pulsation and on mechanical speed ($R8_s$). The stator angle is obtained by integrating the stator pulsation ($R10_s$).

3.3.2 Vector Control of the Rotor Flux and of the Torque

3.3.2.1 Model Inversion Control Structure

The COG in Fig. 3.20 represents all of the previously established relations, which are useful for the modeling of the machine if the rotor flux is correctly oriented.

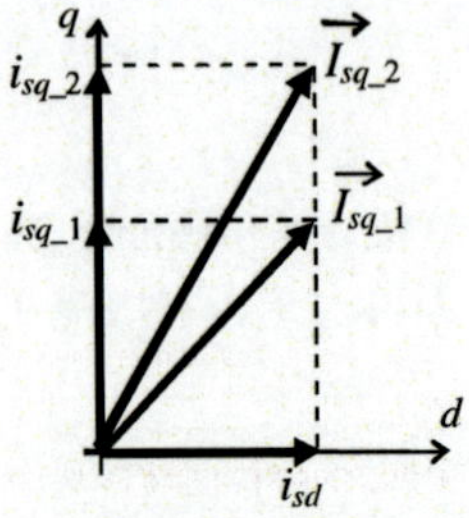

Fig. 3.21 Effect of the component in quadrature on the stator current vector

To control the torque, the variables that need to be set are obviously the magnetizing current $i_{\phi r}$ and the current i_{sq}, but also the instantaneous speed ω_r, and/or the angle θ_r, so that the condition on the orientation of the reference frame is validated. The graph shows that each Park component of the stator current enables the separate regulation of rotor flux and generated torque. It seems logical to affect the setting of $i_{\phi r}$ to i_{sd}, and the setting of the torque T to i_{sq}. This way, in the case of a constant magnetizing-current mechanism, the flux direct component is also constant. The obtained torque value is then directly proportional to the quadrature component of the stator current (Fig. 3.21). This allows for a better energetic optimization because the stator current is obtained combining these two components. Therefore, it is not proportional to the desired torque.

3.3.2.2 Torque Control Through Model Inversion

If the magnetizing current is controlled in order to remain constant at a given value, the torque becomes proportional to the quadrature component of the stator current. By applying the inversion principles to the process in gray on Fig. 3.20, the torque control at a reference value (T_{ref}) is then based on (Fig. 3.22):

- a relation obtained by direct inversion of the $R12$ instantaneous relation, which makes it possible to determine the reference of the stator current quadrature component

$$\boxed{R12c \rightarrow i_{sq\_reg} = \frac{1}{p(1-\sigma)\,L_s}\,\frac{1}{i_{\phi r\_ref}}T_{ref}} \tag{3.59}$$

- a feedback of this current, derived from the direct inversion of the $R13$ causal relation

$$\boxed{R13c \rightarrow u_{sq\_reg} = C_q\left(i_{sq\_reg} - \widehat{i}_{sq}\right)} \tag{3.60}$$

C_q corresponds to the controller dedicated to the feedback of the current;
- a compensation of the stator e.m.f.

$$\boxed{R14c \rightarrow v_{sq\_reg} = u_{sq\_reg} + \tilde{e}_{sq}} \tag{3.61}$$

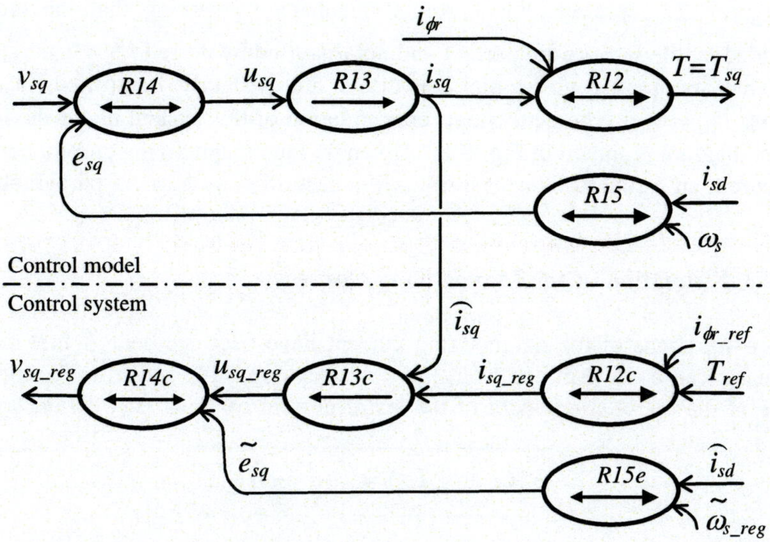

Fig. 3.22 Determination of the control quadrature voltage

- an estimation of the quadrature component of the stator e.m.f. through an observation relation

$$\boxed{R15e \rightarrow \tilde{e}_{\text{sq}}(s) = L_s^* \frac{1+\sigma\tau_r^* s}{1+\tau_r^* s} .\mathbf{L}\left(\widehat{i}_{\text{sd}}(t) \cdot \tilde{\omega}_{s\_\text{reg}}(t)\right)} \qquad (3.62)$$

The symbol * means that the corresponding variable has been previously measured but is likely to vary during operation. A sensitivity study in relation to the variations of these parameters will be carried out in the next chapter.

The speed $\tilde{\omega}_{s\_\text{reg}}$ and the current are expressed according to time. The e.m.f.s appear as nonlinear relations, of which the Laplace transform (written *L*) is a convolution product. If one of the two values is slower than the other one, this relation can be simplified and becomes linear.

The compensation (Eq. 3.61) is only possible if the e.m.f. varies more slowly than the current to be controlled. Given that this e.m.f. depends on the stator pulsation (the speed of the frame of reference linked to the stator), this condition is verified. The compensation simultaneously decouples and linearizes the system. Indeed, if $\tilde{e}_{\text{sq}} \cong e_{\text{sq}}$ and if $v_{\text{sq}} = v_{\text{sq\_reg}}$, this sub-system is not disturbed and becomes independent. Under those conditions, the choice of the control parameters for the controllers only depends on the dynamic wanted and on the performances of the motor power supply. It is also advisable to check that the switching frequency of the converter that supplies the machine is not both a limit and a choice criterion for the dynamic (the response time must be in the order of five switching periods).

Figure 3.22 represents the currents' feedbacks, assuming that the measures provided by the sensors are perfect and not distorted by noise ($\widehat{i}_{sd} = i_{sd}, \widehat{i}_{sq} = i_{sq}$) and that the power electronic converters are perfectly controlled ($v_{sd\_reg} = v_{sd}, v_{sq\_reg} = v_{sq}$). The equivalent representation of this system under the form of a bloc diagram is shown in Fig. 3.26. Given that the magnetizing current cannot be measured, an estimator is necessary and is described in the next paragraph.

3.3.2.3 Estimation of the Magnetizing Current

Two expressions of the magnetizing current have been studied. A first dynamic estimator can be determined from Eq. 3.43 (relation R16) and from the measured value of the direct component of the stator current by (Vas 1990) (Fig. 3.23):

$$\boxed{R16e \rightarrow \tilde{i}_{\phi r}(s) = \frac{1}{1 + \tau_r^* s}\,\widehat{i}_{sd}(s)} \qquad (3.63)$$

This passage shows the importance of precision when measuring the rotor time constant.

Thus, temperature and magnetic saturation are likely to spoil the estimation. The sensitivity of this time constant on the control law and so its robustness will be studied in Chap. 4.

A first feedback control law of the magnetizing current makes it possible to determine the direct component of the stator current (control relation *R*16*c*) using a controller. A second feedback control law then allows for the ascertaining of the direct component of the stator voltage to be adjusted (control relation: *R*17c) using a C_{isd} controller.

A bypass of the stator e.m.f. linked to the quadrature axis is also carried out:

$$\boxed{R18c \rightarrow v_{sd\_reg} = u_{sd\_reg} + \tilde{e}_{sd}} \qquad (3.64)$$

With this objective, an estimation of this component is necessary:

$$\boxed{R19e \rightarrow \tilde{e}_{sd} = -\sigma^* L_s^* \tilde{\omega}_{s\_reg} \widehat{i}_{sq}} \qquad (3.65)$$

Using the second expression of the magnetizing current (Eq. 3.49), a second estimator can also be used:

$$\boxed{R20e \rightarrow \tilde{i}_{\phi r}(s) = \frac{1}{R_s^*} \frac{1}{\sigma \tau_s^* \tau_r^* s^2 + (\tau_s^* + \tau_r^*) s + 1} u_{sd\_reg}(s)} \qquad (3.66)$$

This second estimator requires that both time constants (rotor and stator) are very accurate. Moreover, the control law must then be modified (Fig. 3.24).

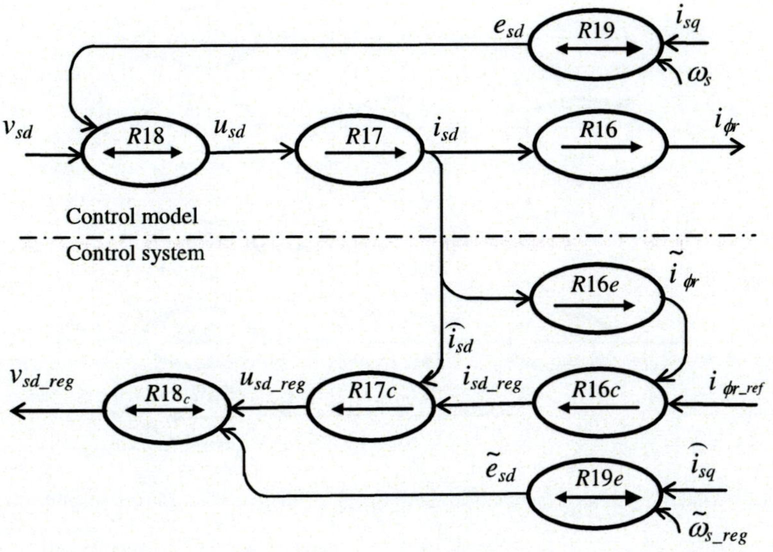

Fig. 3.23 Determination of the stator direct control voltage with the help of the first estimator

3.3.2.4 Explicit Technique for Frame Orientation

The control with a priori on the orientation of the rotor flux consists in calculating the orientation angles (Fig. 2.8), assuming that the quadrature component of the rotor flux is equal to zero. This feed-forward technique is also known as indirect vector control. To cancel the rotor-flux component of axis q, the Park frame of reference must be oriented so that axis d and the rotor-flux vector are aligned. The orientation to be given imposes the cancelation of one of the e.m.f.s (relation $R6_{\text{rd}}$, Table 3.4); the other e.m.f. (relation $R21$) then represents a condition on the rotor pulsation. The rotor pulsation can be estimated by using the Eq. 3.58. The two rotor-flux estimators previously presented can be used here. However, they propagate the signal noise of the stator current.

Assuming a perfect regulation of the direct component of the rotor flux or of the magnetizing flux, those variables can be replaced by their reference value. In this case, the quadrature component of the stator current becomes the only control quantity:

$$\boxed{R21e \rightarrow \tilde{\omega}_{r\_\text{reg}} = \frac{R_r^*}{L_r^*}\frac{\hat{i}_{\text{sq}}}{i_{\phi r\_\text{ref}}}} \tag{3.67}$$

The angle is then obtained through the integration of the corresponding rotor pulsation (Fig. 3.25):

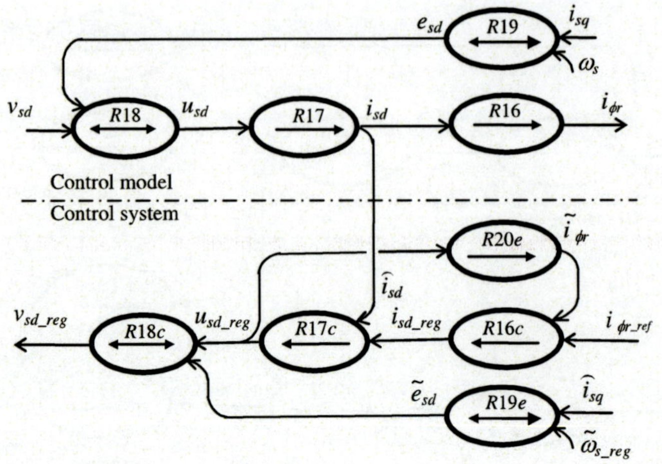

Fig. 3.24 Determination of the stator direct voltage of the control with the second estimator

$$\boxed{R11e \rightarrow \tilde{\theta}_{r\_\text{reg}} = \int_{t_0}^{t_0+\Delta t} \tilde{\omega}_{r\_\text{reg}} \mathrm{d}t + \tilde{\theta}_{r\_\text{reg}}(t_0)} \tag{3.68}$$

It must be noted that the estimation of the angle is particularly sensitive to the rotor time constant, which depends, among others, on rotor temperature, magnetic saturation level, and skin effect due to the high frequencies induced to the rotor by the voltage inverter. The frame orientation technique makes it possible to express stator pulsation depending on rotor pulsation and speed (Fig. 2.8, Eq. 2.85):

$$\boxed{R10e \rightarrow \tilde{\theta}_{s\_\text{reg}} = \hat{\theta} + \tilde{\theta}_{r\_\text{reg}}} \tag{3.69}$$

It is then necessary to provide the machine with an incremental coder that will allow measuring the position of the machine shaft ($\hat{\theta}$) with position encoder. This orientation technique (Fig. 3.25) explicitly uses the speed of the Park reference frame in the rotor frame as control quantity. The stator pulsation necessary to the relations ($R15e$ and $R21e$) is obtained by using (Fig. 3.26):

$$\boxed{R8e \rightarrow \tilde{\omega}_{s\_\text{reg}} = p\hat{\Omega} + \tilde{\omega}_{r\_\text{reg}}} \tag{3.70}$$

3.3.2.5 Implicit Technique for Frame Orientation

From relation $R6_{\text{sq}}$ (Table 3.6) an estimation of the stator pulsation can be obtained with relation $R1'_{\text{sq}}$:

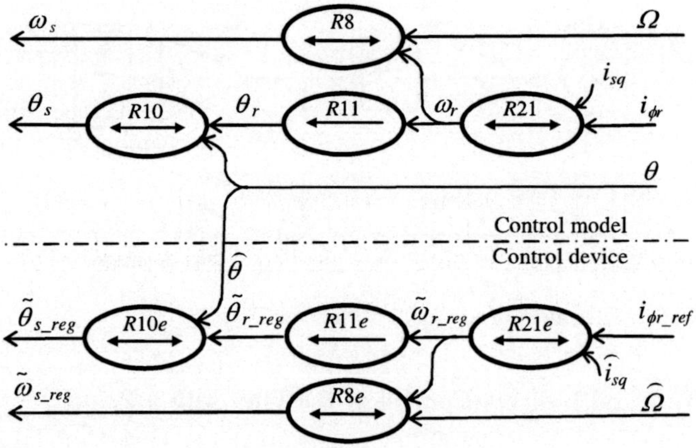

Fig. 3.25 COG of the model and of the explicit orientation

$$\tilde{\omega}_{s\_reg} = \frac{\tilde{e}_{sq}}{\tilde{\phi}_{sd}} = \frac{v_{sq\_reg} - R_s^* \widehat{i}_{sq} - s\tilde{\phi}_{sq}}{\tilde{\phi}_{sd}} \tag{3.71}$$

Or with relation $R3'_{sq}$ to estimate the quadrature component of the stator flux and relation $R3_{sd}$ to estimate the direct component of the stator flux:

$$\tilde{\omega}_{s\_reg} = \frac{v_{sq\_reg} - R_s^* \widehat{i}_{sq} - s\sigma^* L_s^* \widehat{i}_{sq}}{L_s^* \widehat{i}_{sd} + M^* \tilde{i}_{rd}} \tag{3.72}$$

The direct component of the stator current is calculated with relation $R3_{rd}$:

$$\tilde{\omega}_{s\_reg} = \frac{v_{sq\_reg} - \left(R_s^* + \sigma^* L_s^* s\right) \widehat{i}_{sq}}{\frac{L_s^*}{M^*}\left(\tilde{\phi}_{rd} - \sigma^* L_r^* \tilde{i}_{rd}\right)} \tag{3.73}$$

Under steady conditions, the rotor flux is constant and the direct component of the rotor current is equal to zero (relation $R1'_{rd}$), hence:

$$\boxed{\tilde{\omega}_{s\_reg} = \frac{v_{sq\_reg} - \left(R_s^* + \sigma^* L_s^* s\right) \widehat{i}_{sq}}{L_s^* \tilde{i}_{\phi r}}} \tag{3.74}$$

In this formula, we can also introduce a predictive action using directly the frame of the magnetizing current ($i_{\varphi r\_ref}$) rather than its estimated value. This procedure provides an *implicit orientation of the frame of reference*, since knowing ω_r is not necessary as in the case previously studied.

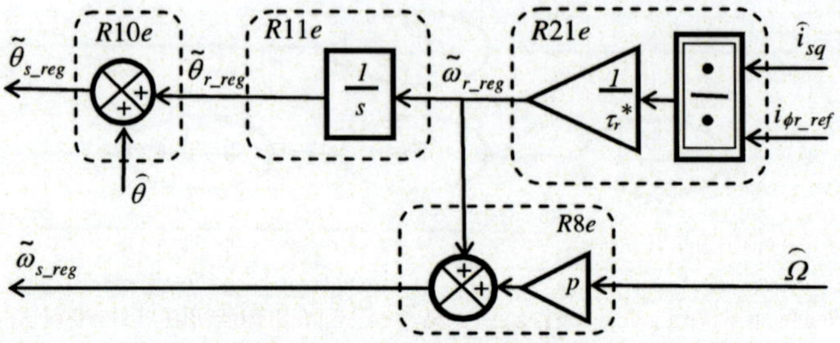

Fig. 3.26 Explicit orientation of the frame with block diagrams

3.3.2.6 Direct and Indirect Control of the Flux with a Priori

Figure 3.27 is a global representation of the organization of the different functions of the control with a priori orientation of the rotor flux. The obtained control structure exploits the natural time decoupling that exists between the magnetizing-current time constant and the stator-current time constant as well as between the mechanical time constant and the rotor flux constant. The very fast control dynamic of the stator currents allows the overlapping of those loops within the control structure. A decoupling by estimation of the e.m.f. is also carried out. This structure relies on the use of a magnetizing current regulation (controller $C_{\phi}(s)$) for which an estimation has been obtained directly from the equations previously simplified with the assumed orientation of the rotor flux. This control structure is also called *direct vector control* because the flux is regulated.

Under steady state conditions, the magnetizing current becomes equal to the direct component of the stator current (Eq. 3.43). This way, a second control structure, called *indirect vector control*, is also used and does not apply flux control. The rest of the control device remains the same (Fig. 3.27).

The reference torque generally comes from a closed-loop control of the speed (Fig. 3.28). Transforms must be applied to the measures of the stator currents; inverse transforms must be applied to the stator control voltages to determine the corresponding three-phase voltages that will be delivered by an inverter (Hautier and Caron 1999).

3.4 Rotor-Flux Observer-Based Control

3.4.1 Control Without a Priori on the Flux Orientation

Rotor flux is of major importance within the torque control device, because it contributes to the control of the electromagnetic torque (equation *R*12*c*, Fig. 3.22), acts directly on the indirect orientation of the Park frame of reference (Fig. 3.24) and, by its influence on the calculation of the stator pulsation, participates to the estimation of the e.m.f. for axis decoupling (equation *R*19*e* of Fig. 3.27). From an

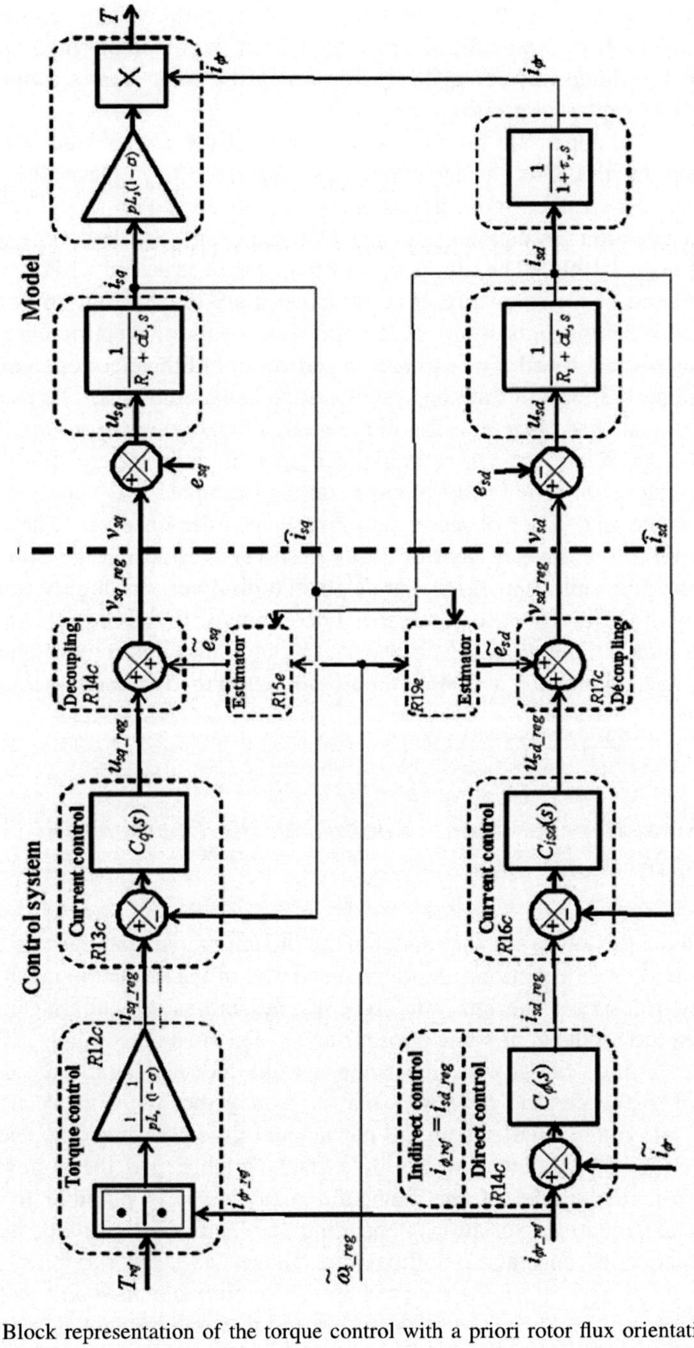

Fig. 3.27 Block representation of the torque control with a priori rotor flux orientation

economic point of view, the use of a flux sensor is not advisable. In the previous control structure, the rotor flux is estimated from the available measures and by using the simplified physic equations for which the rotor flux is supposed to be well-oriented (open-loop estimator).

In order to compensate for the consequences of the unavoidable errors on the calculation of this flux, it is sometimes necessary to replace this estimator (sometimes called open-loop estimator) by an observer (Sect. 3.1.7.3) that will allow a correction by counter reaction (Verghese and Sanders 1988—Piertzak-David of Fornel 1994). The aim is to obtain a control structure without a priori on the orientation of the rotor flux. It is often necessary to carry out an estimation of the two flux components when developing estimators of electromagnetic torque and of mechanical speed. The estimation and control of the electromagnetic torque play an important part in traction systems, such as locomotives. The estimation of the mechanical speed is interesting in the more and more numerous applications in which the use of a speed sensor is avoided, as well as in supervision systems.

In this part, we will be developing, as an example, two classical rotor-flux observers: reduced order observer and complete order observer. The main difficulty linked to the development of a flux observer is choosing the observer gains that set the poles and thus the dynamics of the observer, but highly influence the sensitivity of the observer to parametric uncertainties. In the next chapter, we will be developing a theoretical sensitivity study that allows determining gains for the observer that constitute a good compromise among dynamics, sensitivity and simplicity.

3.4.2 State Representation Adapted to the Observation of the Rotor Flux

To develop a flux observer, the model of the induction machine must be transposed in the form of state equations. A priori, the model of the induction machine cannot be easily transposed because the machine equations are nonlinear. In those equations, the products of some currents and speed are state variables (Eq. 2.121). However, in most cases, we can assume that the mechanical speed varies slowly compared to the electric quantities that is, to say that the highest electric time constants are much smaller than the mechanical time constant. The speed is then considered as a parameter and not as a state variable, and the system becomes linear. From the model of the induction machine, it is possible to determine various state-equations variations depending on whether the currents, the fluxes or a combination of currents and fluxes are chosen as state variables. It is most common to try and determine the rotor flux. The components d and q of this rotor flux are then chosen as state variables. For the two remaining state variables, the components d and q of the stator current are selected because this current can easily be measured in practice.

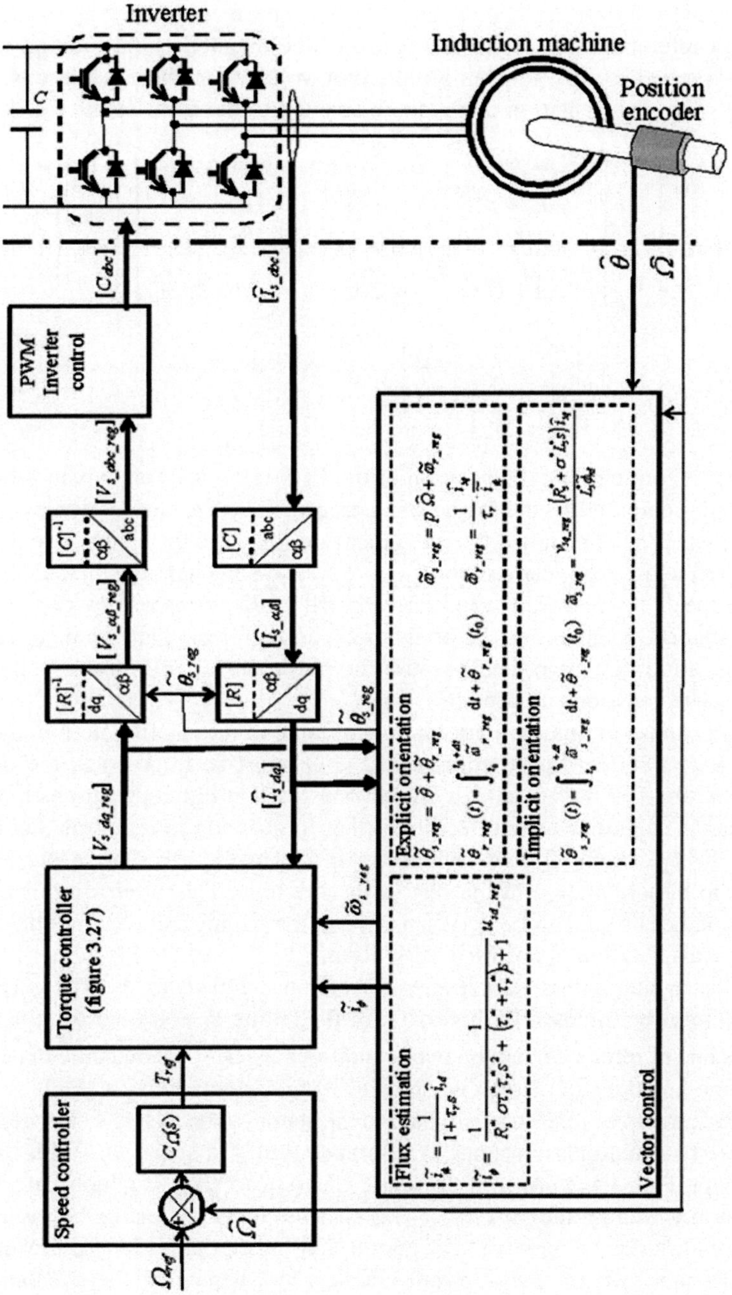

Fig. 3.28 Structure of the control system for vector control with a priori rotor-flux orientation

The model shown in Chap. 2 (Eq. 2.121) has been established by synchronizing the Park reference frame with the vector of the magnetic field rotating in the air gap at a speed ω_s. By imposing null rotor voltages in those equations, the following state representation is obtained, adapted to the form of the Eq. 3.28:

$$\begin{bmatrix} \dot{\phi}_{\mathrm{rd}} \\ \dot{\phi}_{\mathrm{rq}} \\ \dot{i}_{\mathrm{sd}} \\ \dot{i}_{\mathrm{sq}} \end{bmatrix} = \begin{bmatrix} -\frac{R_r}{L_r} & (\omega_p - p\Omega) & \frac{MR_r}{L_r} & 0 \\ -(\omega_p - p\Omega) & -\frac{R_r}{L_r} & 0 & \frac{MR_r}{L_r} \\ \frac{MR_r}{\sigma L_s L_r^2} & \frac{M}{\sigma L_s L_r} p\Omega & -\frac{R_{\mathrm{sr}}}{\sigma L_s} & \omega_p \\ -\frac{M}{\sigma L_s L_r} p\Omega & \frac{MR_r}{\sigma L_s L_r^2} & -\omega_p & -\frac{R_{\mathrm{sr}}}{\sigma L_s} \end{bmatrix} \begin{bmatrix} \phi_{\mathrm{rd}} \\ \phi_{\mathrm{rq}} \\ i_{\mathrm{sd}} \\ i_{\mathrm{sq}} \end{bmatrix} + \begin{bmatrix} 0 & 0 \\ 0 & 0 \\ \frac{1}{\sigma L_s} & 0 \\ 0 & \frac{1}{\sigma L_s} \end{bmatrix} \begin{bmatrix} v_{\mathrm{sd}} \\ v_{\mathrm{sq}} \end{bmatrix} \tag{3.75}$$

For this formulation, the position of the Park reference frame is indifferent and its speed, compared to the frame of reference linked to the motionless stator, is written ω_p. It must be noted that the quantities linked to the stator have a speed ω_s compared to the motionless stator (Sect. 2.4.2.1). The quantities linked to the rotor have a speed ω_s compared to the rotor, which itself rotates at a speed Ω compared to the stator. Depending on the orientation chosen for the Park frame of reference, state equations are simplified because the pulsation ω_p will have specific values. Three cases must be considered.

If we choose to link the Park reference frame to the rotating field in the air gap (as in Sect. 2.4.2), a speed equal to $\omega_p = \omega_s$ must be imposed in the Eqs. 3.75. This solution has the advantage of showing that the different variables evolve as continuous quantities, which facilitates their processing in real time. On the other hand, the pulsations ω_s and ω_r interact in the matrix $[A]$. These pulsations are linked to the orientation of the flux in the machine. This orientation is a delicate operation because it can be very sensitive to the parametric uncertainties. For this reason, this solution is not used very often.

To set a motionless Park frame of reference linked to the stator (Fig. 2.8), $\omega_p = 0$ must be imposed in the Eq. 3.75. By setting $\theta_s = 0$ (and so $\theta_r = -\theta$), the Park frame of reference and the two-axe reference $(\vec{O}_\alpha, \vec{O}_\beta)$ are coincident (Fig. 2.8) and the notations (d, q) can be substituted to the notations (α, β). It is said that "the Park frame of reference is linked to the stator". The stator vectors represented in those two motionless frames of reference turn at a speed ω_s. With this orientation, the Park transform (on the stator quantities $P(\theta_s)$) is simplified to a linear transform without rotation (Eq. 2.83). This solution is interesting because the only pulsation playing a part in the matrix $[A]$ is the pulsation ω linked to the mechanical speed. The pulsations ω_s and ω_r which depend on the orientation of the flux do not intervene any more. Nevertheless, it can be objected to this solution that the flux variables, currents and voltages are sinusoidal quantities with a pulsation ω_s. In spite of this inconvenience, this solution is very often chosen.

To link the Park reference frame to the axis speed (Fig. 2.8), $\theta_r = 0$ (and so $\theta_s = 0$) must be imposed and, under those conditions, $\omega_p = p\Omega$ must be set in the Eq. 3.75. In this rotating frame of reference, the rotor quantities become constant. Again, the only pulsation playing a part in the matrix $[A]$ is ω. It must be noticed that, in order to bring back the components of the stator current in the frame of reference linked to the rotor and rotating with it, a Park transform is necessary to make a rotation of the position angle of the rotor, which is generally tapped by a coder. This solution is hardly used, except to perform a parametric identification or to estimate the magnetic states of the machine.

Given that the equation system (3.75) is valid in any reference frame, it is not possible to differentiate the index from the variables depending on the frame of reference, which is why we have chosen to use the indexes d and q in all cases.

In all cases, the magnitude of the estimated rotor flux is deduced from the following relation:

$$\left|\tilde{\phi}_r\right| = \sqrt{\tilde{\phi}_{\mathrm{rd}}^2 + \tilde{\phi}_{\mathrm{rq}}^2} \tag{3.76}$$

On the other hand, the orientation of the flux depends on the reference frame in which the observer is developed.

3.4.3 Rotor Flux Observers

3.4.3.1 Reduced Order Observer

To estimate the two components of the rotor flux, the simplest flux dynamic estimator can be deduced from the first two equations of the system (3.75). It is a second-order estimator (Verghese and Sanders 1988; Pietrzak David and de Fornel 1994). The obtained equation is:

$$\begin{bmatrix} \dot{\tilde{\phi}}_{\mathrm{rd}} \\ \dot{\tilde{\phi}}_{\mathrm{rq}} \end{bmatrix} = \begin{bmatrix} -\frac{R_r^*}{L_r^*} & \left(\omega_p - p\hat{\Omega}\right) \\ -\left(\omega_p - p\hat{\Omega}\right) & -\frac{R_r^*}{L_r^*} \end{bmatrix} \begin{bmatrix} \tilde{\phi}_{\mathrm{rd}} \\ \tilde{\phi}_{\mathrm{rq}} \end{bmatrix} + \begin{bmatrix} \frac{M^* R_r^*}{L_r^*} & 0 \\ 0 & \frac{M^* R_r^*}{L_r^*} \end{bmatrix} \begin{bmatrix} \hat{i}_{\mathrm{sd}} \\ \hat{i}_{\mathrm{sq}} \end{bmatrix} \tag{3.77}$$

Note that starred components are estimates. These equations can also be written as:

$$\left[\dot{\tilde{\phi}}_{r\_\mathrm{dq}}\right] = [A_r]\left[\tilde{\phi}_{r\_\mathrm{dq}}\right] + [B_r]\left[\hat{I}_{s\_\mathrm{dq}}\right] \tag{3.78}$$

with $[A_r] = \begin{bmatrix} -\frac{R_r^*}{L_r^*} & \left(\omega_p - p\hat{\Omega}\right) \\ -\left(\omega_p - p\hat{\Omega}\right) & -\frac{R_r^*}{L_r^*} \end{bmatrix}$ and $[B_r] = \frac{M^* R_r^*}{L_r^*}\begin{bmatrix} 1 & 0 \\ 0 & 1 \end{bmatrix}$.

Figure 3.29 shows a graphic representation of this model.

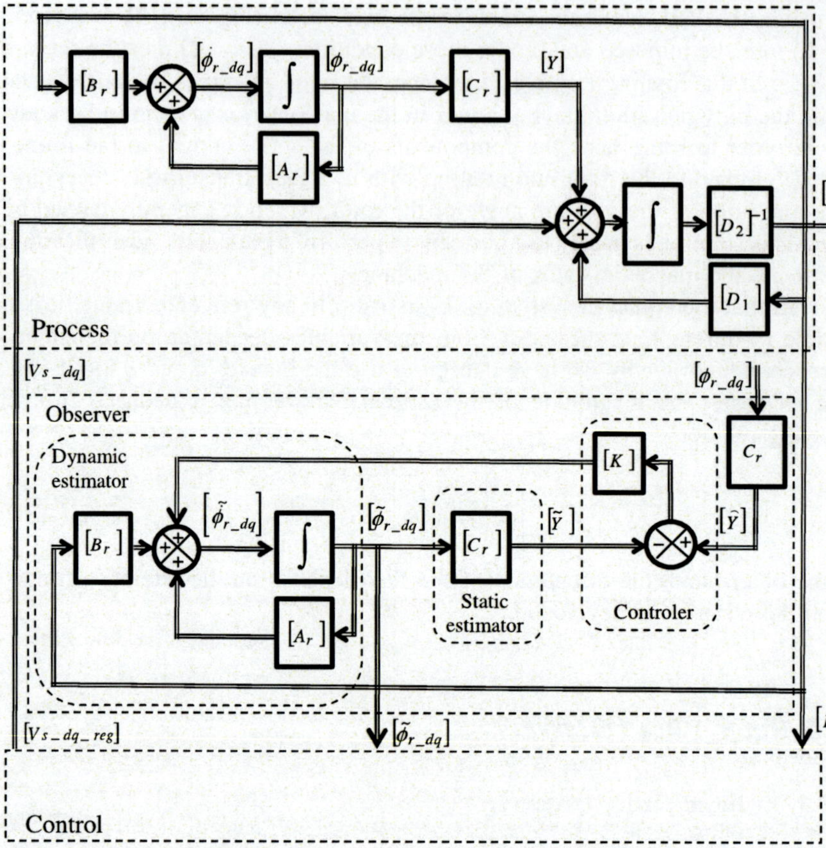

Fig. 3.29 Functional diagram of the reduced-order rotor-flux observer

To obtain an observer from the system (3.75), a corrective term computed from an output vector (estimated and calculated) must be added:

$$[\varepsilon] = [K]\left[\left[\widehat{Y}\right] - \left[\tilde{Y}\right]\right] = [K]\left[\left[\widehat{Y}\right] - [C_r]\left[\tilde{\phi}_{r\_\mathrm{dq}}\right]\right] \tag{3.79}$$

In this study, a symmetrical structure of the gain has been selected to echo the symmetry of the state matrix $[A_r]$:

$$[K] = \begin{bmatrix} -k_1 & k_2 \\ -k_2 & -k_1 \end{bmatrix} \tag{3.80}$$

Consequently, the vector of the calculated outputs which must be taken into account depends on the rotor fluxes (Fig. 3.29):

$$\left[\widehat{Y}\right] = [C_r]\left[\tilde{\phi}_{r\_\mathrm{dq}}\right] \tag{3.81}$$

This term can be determined from the last two equations of the system (3.75). The equations corresponding to stator currents in the equation system (3.75) are written as:

$$[D_2]\left[\dot{I}_{s\_dq}\right] = [C_r]\left[\phi_{r\_dq}\right] + [D_1]\left[I_{s\_dq}\right] + \left[V_{s\_dq}\right] \tag{3.82}$$

with $[D_2] = \sigma^* L_s^* \begin{bmatrix} 1 & 0 \\ 0 & 1 \end{bmatrix}$, $[C_r] = \frac{M^*}{L_r^*}\begin{bmatrix} \frac{R_r^*}{L_r} & p\Omega \\ -p\Omega & \frac{R_r^*}{L_r^*} \end{bmatrix}$ and

$[D_1] = \begin{bmatrix} -R_{sr}^* & \sigma^* L_s^* \omega_p \\ -\sigma^* L_s^* \omega_p & -R_{sr}^* \end{bmatrix}$.

All these values are considered measurable and a new expression of the measured output vector can be determined:

$$\left[\widehat{Y}\right] = [C_r]\left[\widehat{\phi}_{r\_dq}\right] = [D_2]\left[\widehat{\dot{I}}_{s\_dq}\right] - [D_1]\left[\widehat{I}_{s\_dq}\right] - \left[V_{s\_dq\_reg}\right] \tag{3.83}$$

The three equation systems (3.78), (3.79) and (3.83) give the following observer:

$$\begin{aligned}\left[\dot{\tilde{\phi}}_{r\_dq}\right] = {} & [A_r]\left[\tilde{\phi}_{r-dq}\right] + [B_r]\left[\widehat{I}_{s\_dq}\right] \\ & + [K]\left(\left([D_2]\left[\widehat{\dot{I}}_{s\_dq}\right] - [D_1]\left[\widehat{I}_{s\_dq}\right] - \left[V_{s\_dq\_reg}\right] - [C_r]\left[\tilde{\phi}_{r\_dq}\right]\right)\right)\end{aligned} \tag{3.84}$$

Or in a more compact form:

$$\boxed{\left[\dot{\tilde{\phi}}_{r\_dq}\right] = [A_k]\left[\tilde{\phi}_{r\_dq}\right] + [B_k]\left[\widehat{I}_{s\_dq}\right] + [K]\,[D_2]\left[\widehat{\dot{I}}_{s\_dq}\right] - [K]\left[V_{s\_dq\_reg}\right]} \tag{3.85}$$

Matrixes $[A_k]$ and $[B_k]$ take the following forms:

$$[A_k] = [A_r] - [K]\,[C_r] = \begin{bmatrix} a_{1k} & -a_{2k} \\ a_{2k} & a_{1k} \end{bmatrix} \tag{3.86}$$

and

$$[B_k] = [B_r] - [K]\,[D_1] = \begin{bmatrix} b_{1k} & -b_{2k} \\ b_{2k} & b_{1k} \end{bmatrix} \tag{3.87}$$

with

$$a_{1k} = -\frac{R_r^*}{L_r^*} + \frac{M^*}{L_r^*}\left(k_1 \frac{R_r^*}{L_r^*} + k_2\, p\Omega\right) \tag{3.88}$$

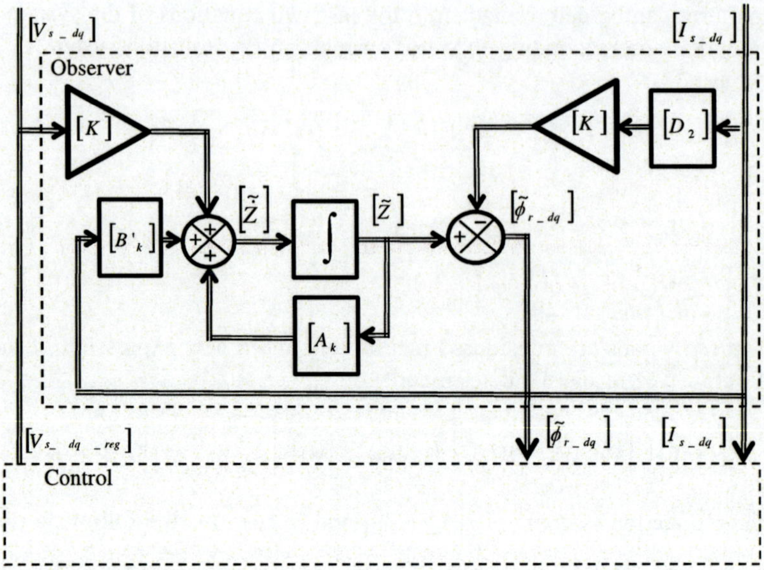

Fig. 3.30 Functional diagram of the reduced-order rotor-flux observer

$$a_{2k} = -(\omega_p - p\Omega) - \frac{M^*}{L_r^*}\left(k_1\, p\Omega - k_2 \frac{R_r^*}{L_r^*}\right) \tag{3.89}$$

$$b_{1k} = M^* \frac{R_r^*}{L_r^*} - k_1 R_{sr}^* + k_2\, \sigma^*\, L_s^* \omega_p \tag{3.90}$$

$$b_{2k} = -k_1 \sigma^* L_S^* \omega_p - k_2 R_{sr}^* \tag{3.91}$$

In order to estimate the flux from the observer deduced from (3.78), it is required to know the current derivative. However, deriving the measured current is not necessary to obtain this value, as a simple variable change in the equation allows for the current derivative without calculation. A new state vector can thus be defined, which takes into account all derived terms (Fig. 3.30):

$$\left[\tilde{Z}\right] = \left[\tilde{\phi}_{r\_\mathrm{dq}}\right] + [K]\,[D_2]\left[\widehat{I}_{s\_\mathrm{dq}}\right] \tag{3.92}$$

Observer (3.85) can then be rewritten as:

$$\left[\dot{\tilde{Z}}\right] = [A_k]\left[\tilde{Z}\right] + \left[B_k'\right]\left[\widehat{I}_{s\_\mathrm{dq}}\right] + [K]\left[V_{s\_\mathrm{dq\_reg}}\right] \tag{3.93}$$

with

$$\left[B_k'\right] = [B_k] - [A_k]\,[K]\,[D_2] \tag{3.94}$$

Finally, the flux vector can be deduced from the following relation:

$$\left[\tilde{\phi}_{r\_\mathrm{dq}}\right] = \left[\tilde{Z}\right] - [K]\,[D_2]\left[\widehat{I}_{s\_\mathrm{dq}}\right] \tag{3.95}$$

Observer (3.85) is called a reduced-order observer as its second order is smaller than the order of the electrical part of the induction machine. The full-order observer presented in the next section is of the same order as the electrical part, i.e. of the fourth order.

3.4.3.2 Full-Order Observer

If the whole equation system (3.75) is taken as estimator, the components of the stator current must be estimated in addition to the components of the rotor flux. The system can be rewritten in a more compact form as (Verghese and Sanders 1988—Pietrzak David and de Fornel 1994):

$$\begin{bmatrix} \left[\dot{\tilde{\phi}}_{r\_\mathrm{dq}}\right] \\ \left[\dot{\tilde{I}}_{s\_\mathrm{dq}}\right] \end{bmatrix} = \begin{bmatrix} [A_{c1}] & [A_{c2}] \\ [A_{c2}] & [A_{c4}] \end{bmatrix} \begin{bmatrix} \left[\tilde{\phi}_{r\_\mathrm{dq}}\right] \\ \left[\tilde{I}_{s\_\mathrm{dq}}\right] \end{bmatrix} + \begin{bmatrix} [0] \\ [B_c] \end{bmatrix} \left[V_{s\_dq\_reg}\right] \tag{3.96}$$

with

$$[A_{c1}]\begin{bmatrix} -\frac{R_r^*}{L_r^*} & -(\omega_p - p\Omega) \\ -(\omega_p - p\Omega) & -\frac{R_r^*}{L_r^*} \end{bmatrix} [A_{c2}] = \frac{M^* R_r^*}{L_r^*}\begin{bmatrix} 1 & 0 \\ 0 & 1 \end{bmatrix} [A_{c3}] = \frac{M^*}{\sigma^* L_S^* L_r^*}\begin{bmatrix} \frac{R_r^*}{L_r^*} & p\Omega \\ -p\Omega & \frac{R_r^*}{L_r^*} \end{bmatrix}$$

$$[A_{c4}] = \begin{bmatrix} \frac{R_{sr}^*}{\sigma^* L_S^*} & \omega_p \\ -\omega_p & -\frac{R_{sr}^*}{\sigma^* L_S^*} \end{bmatrix} [B_{c2}] = \frac{1}{\sigma^* L_S^*}\begin{bmatrix} 1 & 0 \\ 0 & 1 \end{bmatrix}$$

This system must be compared with the general representation (Eq. 3.28), written as:

$$\left[\dot{X}\right] = [A_c]\,[X] + [B_c]\,[U] \tag{3.97}$$

The output quantities taken into account for the design of the observer are the stator currents:

$$\left[\tilde{Y}\right] = \left[\tilde{I}_{s\_dq}\right] = \begin{bmatrix} 0 & 0 & 1 & 0 \\ 0 & 0 & 0 & 1 \end{bmatrix} \begin{bmatrix} \tilde{\phi}_{\mathrm{rd}} \\ \tilde{\phi}_{\mathrm{rq}} \\ \tilde{i}_{\mathrm{sd}} \\ \tilde{i}_{\mathrm{sq}} \end{bmatrix} = [C_c]\left[\tilde{X}\right] \tag{3.98}$$

The full-order observer is obtained by adding a correction term to the estimator (3.96). This term is determined by the difference between the estimated and measured current values:

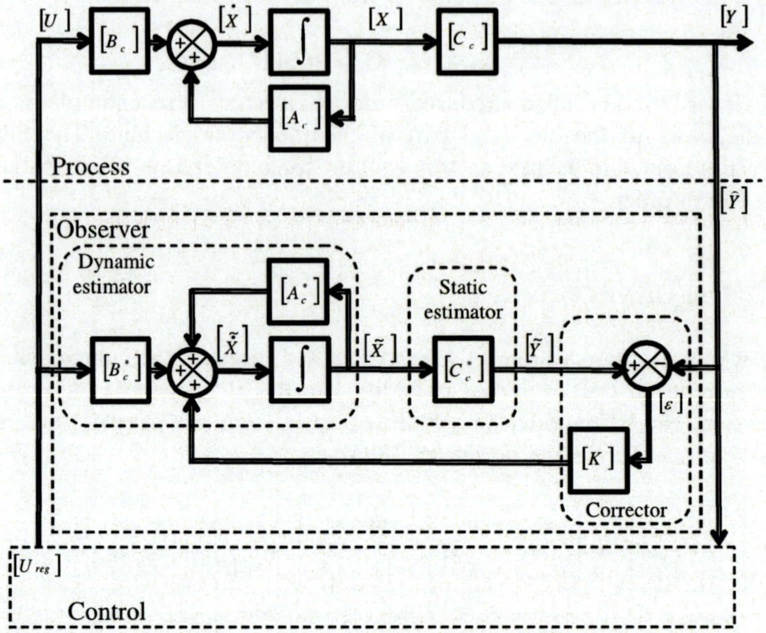

Fig. 3.31 Functional diagram of the full-order rotor-flux observer

$$\begin{bmatrix} \varepsilon_d \\ \varepsilon_q \end{bmatrix} = \left[\hat{Y}\right] - [C_c]\left[\tilde{X}\right] \tag{3.99}$$

The error is mitigated through two gain matrixes to create the following observer:

$$\begin{bmatrix} [K_{12}] \\ [K_{34}] \end{bmatrix} \left(\left[\hat{I}_{s\_dq}\right] - \left[\tilde{I}_{s\_dq}\right] \right) \tag{3.100}$$

with $[K_{12}] = \begin{bmatrix} k_1 & -k_2 \\ k_2 & k_1 \end{bmatrix}$ and

$$[K_{34}] = \begin{bmatrix} k_3 & -k_4 \\ k_4 & k_3 \end{bmatrix} \tag{3.101}$$

By combining mitigated errors (3.100), (3.99) and (3.96), the following result is obtained:

$$\begin{bmatrix} \left[\dot{\tilde{\phi}}_{r\_dq}\right] \\ \left[\dot{\tilde{I}}_{s\_dq}\right] \end{bmatrix} = \begin{bmatrix} [A_{c1}] & [A_{c2}] - [K_{12}] \\ [A_{c3}] & [A_{c4}] - [K_{34}] \end{bmatrix} \begin{bmatrix} \left[\tilde{\phi}_{r\_dq}\right] \\ \left[\tilde{I}_{s\_dq}\right] \end{bmatrix} + \begin{bmatrix} [K_{12}] \\ [K_{34}] \end{bmatrix} \left[\hat{I}_{s\_dq}\right] + \begin{bmatrix} [0] \\ [B_c] \end{bmatrix} \left[V_{s\_dq\_reg}\right] \tag{3.102}$$

The functional diagram of the full-order observer is shown in Fig. 3.31.

3.4.3.3 Gain Determination through Pole Placement

Observer gains must be chosen to set its response time (which must be quicker than the response time of the process) and, at the very least, to guarantee its stability. As a consequence, the poles of the characteristic polynomial of matrix $[A]$ assume specific values, and are from that point dependant on observer gains. With regard to the reduced-order observer defined in Eq. 3.85, the characteristic polynomial of matrix $[A_k]$ can be determined as follows (with $[I]$ the identity matrix):

$$P_c = \det(s[I] - [A_k]) = s^2 - 2a_{1k}s + a_{1k}^2 + a_{2k}^2 \tag{3.103}$$

The general solutions of this polynomial are two complex conjugated poles of the form $(-a + jb)$ and $(-a - jb)$. The characteristic polynomial can be expressed according to the poles:

$$P_c = (s - (-a + jb))\,(s - (-a - jb)) = s^2 + 2as + a^2 + b^2 \tag{3.104}$$

Equating the coefficients of the two polynomials gives:

$$a = -a_{1k} = \frac{R_r^*}{L_r^*} - \frac{M^*}{L_r^*}\left(k_1\frac{R_r^*}{L_r^*} + k_2 p\Omega\right) \tag{3.105}$$

$$b = a_{2k} = -(\omega_p - p\Omega) - \frac{M^*}{L_r^*}\left(k_1 p\Omega - k_2\frac{R_r^*}{L_r^*}\right) \tag{3.106}$$

Gain values can finally be deduced from these relations and expressed, depending on the poles to be imposed:

$$\begin{bmatrix} k_1 \\ k_2 \end{bmatrix} = \frac{1}{\frac{M^*}{L_r^*}\left(\left(\frac{R_r^*}{L_r^*}\right)^2 + p^2\Omega^2\right)} \begin{bmatrix} \left(a - \frac{R_r^*}{L_r^*}\right)\frac{R_r^*}{L_r^*} + (b - (\omega_p - p\Omega))\,p\Omega \\ p\Omega\left(a - \frac{R_r^*}{L_r^*}\right) + \frac{R_r^*}{L_r^*}(-b + (\omega_p - p\Omega)) \end{bmatrix} \tag{3.107}$$

The same method can be used for the full-order observer, to determine the relations between observer gains and poles. However, this observer is characterized by four poles and as many gains, which make calculations more complex. Mathematical software is then used to determine the poles.

Expressions (3.107) show that observer gains depend on mechanical speed (more precisely on the square of the speed). Real-time computation of the gains is therefore necessary in every sampling period. As this calculation is very expensive in CPU time, gains are often set to constant values for some speed ranges. This method, usually empirical, is made even more complex by the fact that gains

Fig. 3.32 Representation of the rotor-flux vector in a static Park reference frame

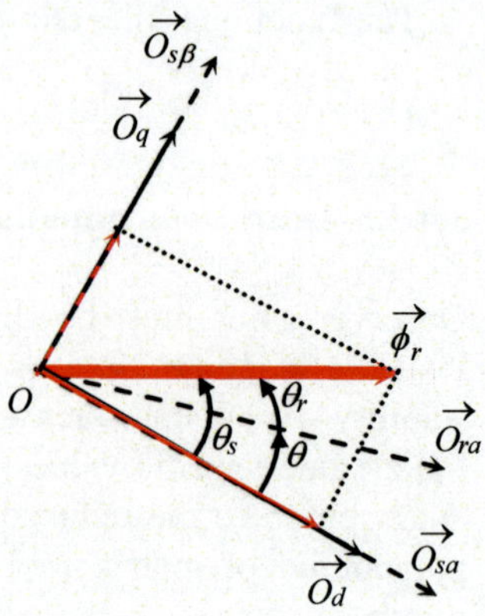

strongly affect observer sensitivity—and thus also control sensitivity—to parameter uncertainties. Experience has shown that, when gains correspond to "good" poles, sensitivity is often higher, and conversely. In the following chapters, we will show that a theoretical sensitivity study makes it possible to define quite easily gain values which make up a good compromise among dynamics, sensitivity and simplicity.

Another solution to choose observer gains is the Kalman filter (Ben Ammar 1993; Piertzak-David and de Fornel 1994; Grellet and Clerc 1997). This method allows for observer dynamics adjustment while filtering state variables. Nevertheless, the observer remains very sensitive to parameter uncertainties.

The extended Kalman filter needs to be considered to reduce this sensitivity. In this observer, parameters are treated as state variables, and as many state equations are added to the initial equation system as there are parameters to be taken into account. In our case, to evaluate variations on stator and rotor resistances and on inductance M, three state equations must be added to the system (see Eq. 3.75). The final observer is of the seventh order, and will therefore be difficult to implement in real-time—especially because, with low electrical time constants, the observer computation time cannot exceed hundreds of microseconds.

3.4.4 Vector Orientation in a Static Reference Frame

The rotor flux vector has a speed ω_r compared to the rotor; the rotor itself is turning at speed Ω compared to a static Park reference frame linked to the stator. In the same reference frame, the rotor flux vector is thus turning at speed ω_s. When the rotor flux is estimated from an observer developed in this reference frame, the angle $\theta_s$can thus be calculated on the basis of projections of the rotor flux vector onto a reference frame $(O, \vec{O}_{s\alpha}, \vec{O}_{s\beta})$ (see Fig. 3.32):

$$\theta_{s\_reg} = \mathrm{arctg}\left(\frac{\tilde{\phi}_{r\beta}}{\tilde{\phi}_{r\alpha}}\right) = \mathrm{arctg}\left(\frac{\tilde{\phi}_{rq}}{\tilde{\phi}_{rd}}\right)_{\text{in the stator frame}} \quad (3.108)$$

The sine and cosine of angle $\theta_s$can be directly deduced from the two components of the estimated flux using Eq. 3.76:

$$\cos(\theta_s) = \frac{\tilde{\phi}_{r\alpha}}{|\tilde{\phi}_r|} = \frac{\tilde{\phi}_{rd}}{|\tilde{\phi}_r|} \text{ and } \sin(\theta_s) = \frac{\tilde{\phi}_{r\beta}}{|\tilde{\phi}_r|} = \frac{\tilde{\phi}_{rq}}{|\tilde{\phi}_r|} \quad (3.109)$$

In these equations, $\tilde{\phi}_{rd}$ and $\tilde{\phi}_{rd}$ correspond to the projections of the rotor flux vectors onto the stator reference frame. Since stator pulsation ω_s is not directly computed, the decoupling terms ($\tilde{e}_{sd}$ and $\tilde{e}_{sq}$, see Fig. 3.27) dependent on ω_s are sometimes omitted. However, these terms can be determined by estimating ω_s with Eq. 3.74.

3.4.5 Direct Vector Control of the Flux Without a Priori Orientation

The structure of the torque control is identical to that of a priori control.

Direct vector control of the flux is obtained through closed-loop control of the magnetizing current, which is not measured but observed with a reduced-order or full-order rotor-flux observer (see Figs. 3.30 and 3.31). Control structure without a priori on rotor-flux orientation must therefore be modified to take into account the observer. An example with an orientation in a static Park reference frame linked to the stator (thus coincident with the frame of reference $\vec{O}_{s\alpha}, \vec{O}_{s\beta}$) is shown in Fig. 3.33.

3.5 Discretization of Estimators and Observers

The observers presented in the previous sections make up a set of differential equations. To integrate them into a microprocessor, their equations must be discretized to give a set of recurrent equations. After the discretization, equation systems (3.31) and (3.30) take the following form (Borne et al. 1993—Aström and Wittenmark 1984):

$$[X(k+1)] = [\phi]\,[X(k)] + [\Gamma]\,[U(k)] \quad (3.110)$$

$$[Y(k)] = [C]\,[X(k)] + [D]\,[U(k)] \quad (3.111)$$

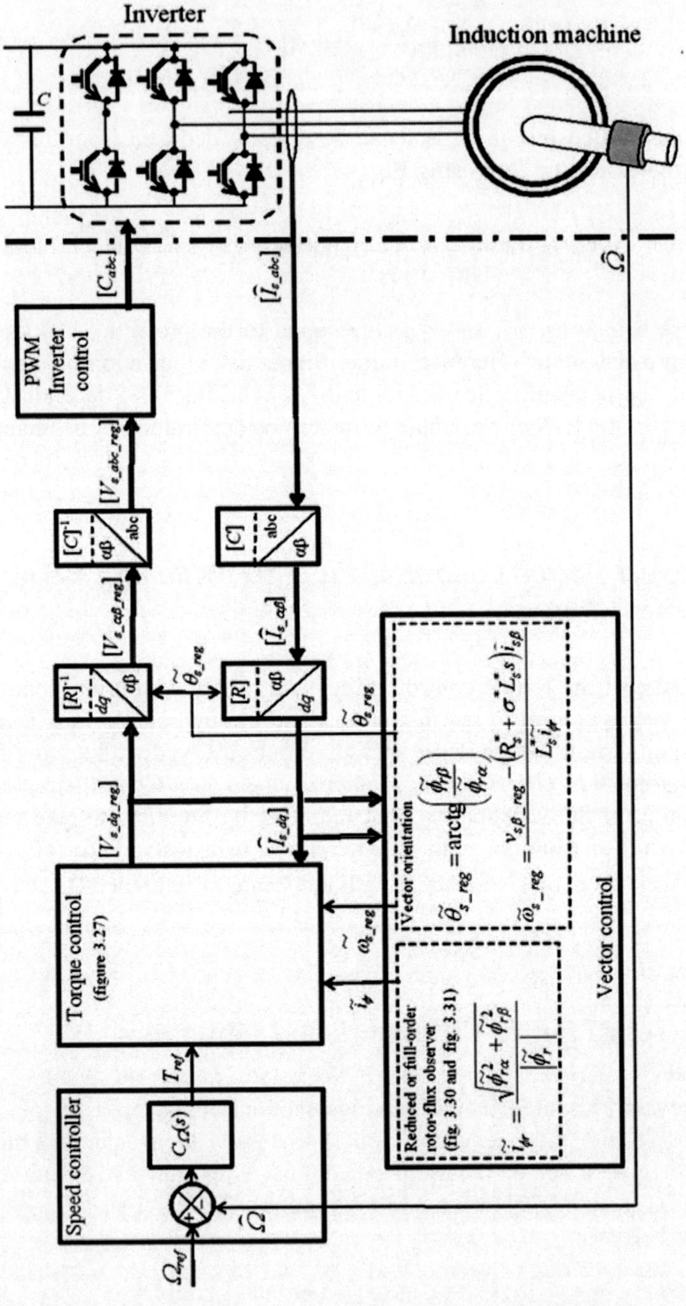

Fig. 3.33 Organization of the control apparatus through vector control without a priori on the rotor-flux orientation

Matrixes $[\Phi]$ and $[\Gamma]$ can be deduced from matrixes $[A]$ and $[B]$ with the introduction of sampling period h:Either through exact calculation :

$$[\phi] = \left[e^{A.h}\right] = L^{-1}\left(s[I] - [A]\right)^{-1} \quad (3.112)$$

$$[\Gamma] = \int_0^h \left[e^{A_k.t}\right] \mathrm{d}t\,[B] \quad (3.113)$$

Or through an approximate computation, obtained by limited development of the matrix exponential:

$$[\Phi] = \left[e^{Ah}\right] \approx [I] + [A]\,h + [A]^2\frac{h^2}{2} + \ldots \quad (3.114)$$

$$[\Gamma] \approx \int_0^h \left([I] + [A]\,t + [A]^2\frac{t^2}{2} + \cdots\right)\mathrm{d}t\,[B] \quad (3.115)$$

Since in our case, matrix $[A]$ must be recalculated at every sampling period and can be of high rank, an approximate computation is often used (see 3.114). Note that the order of the series expansion and the value of the sampling period h both influence the quality of the estimate. However, errors on the flux estimate can be made negligible thanks to a second-order or higher series expansion and a short sampling period (maximum 500 μs).

We have shown that flux estimators can be calculated if pulsation (ω) is considered as a parameter. In practice, mechanical speed varies, so that, for a numerical implementation, the value of ω must be updated at every sampling period. If the separation of the electrical and mechanical modes is not possible, i.e. if the speed is not slowly varying compared to the dynamics of the state quantities, the mechanical equation must then be added to the observed system. In this case, determinist nonlinear observers or extended Kalman filters are called for (Piertzak-David and de Fornel 1994; Von Westerholt 1994). As this case is quite uncommon, it will not be considered in our study.

3.6 Conclusion

Designing a control system requires a precise modeling of the machine to highlight the essential properties which make it possible to determine the control laws adapted to torque control. Depending on the mathematical expression used to model the torque, several strategies exist for the design of induction machine control. Torque expressions can be simplified by changing the orientation of the reference frame used to model the dynamic equations of the machine.

We have introduced a strategy for vector control based on rotor flux orientation. Such an orientation can be implemented either explicitly, by using the equations of the induction machine, or implicitly, through the use of speed conditions linking the various electrical quantities.

The equations obtained in the reference frame show that a precise control of the torque with set dynamic performances requires a vector control of the flux decoupled from the torque. Flux control thus enables the delivery of the maximum torque at all points of the speed range. A wider operating range is thus made possible for low speeds or even for a speed equal to zero (the torque is maintained even when the machine is not in operation).

Moreover, the performance of the inverter is improved through better adjustment to the various operating conditions. A more precise energy control of the machine is then possible, resulting in lower electrical consumption and savings on mechanical maintenance costs of the load driven by industrial application. As an example, these control structures result in the lowest no-load consumption possible, limiting temperature and minimizing operating costs.

Since measuring the flux is not cost-effective, its calculation is essential to guarantee good performance. The equations of the model can be used to create an open-loop estimator, but they do not enable corrections in the event of parameter errors, which is why observers have been developed. Examples of the design of a reduced-order rotor-flux observer and of a full-order observer have been presented. These flux estimators and observers proposed here make up a representative though not exhaustive sample of the possible flux observers and estimators.

The implementation of these control and observation algorithms requires powerful CPUs so the computing time remains low compared to the sampling period.

This chapter has shown that a number of control structures can be used and that their complexity varies depending on additional choices. Another major difference between these control techniques is their sensitivity to parameter variations, which is the object of the next chapter.

References

Aström W. (1984). Computer controlled systems theory and design. Prentice Hall editions, ISBN 978-0-13-314899-2.

Barre P.J., Caron J.P., Hautier J.P., & Legrand M. (1995) Analyse er Modèles, collection Systèmes automatiques, vol. 1, Editions Ellipses, Paris 1995, ISBN 2-7298-5515-7.

Ben Ammar F. (1993). Variateur de vitesse hautes performances pour machine asynchrone de grande puissance. PhD Thesus report from INP Toulouse.

Borne, P., Dauphin-Tanguy, G., Richard, J. P., Rotella, F., & Zambettakis, I. (1990). *Commande et optimisation des processus*. Paris: Editions Technip.

Borne, P., Dauphin-Tanguy, G., Richard, J.P., Rotella, F., & Zambettakis, I. (1993). Analyse et régulation des processus industriels, tome 2, chap. 5, Editions Technip, Paris, ISBN 2-7108-0643-6.

Degobert, P. (1997). Formalisme pour la commande des machines électriques alimentées par convertisseurs statiques, application à la commande numérique d'un ensemble machine asynchrone - commutateur de courant, PhD Thesus report from Université des Sciences et Technologies de Lille.

Grellet, G., & Clerc, G. (1997). Actionneurs électriques, Paris: Eyrolles, ISBN 2-212-09352-7.

Hautier, J.P., & Caron, J.P. (1997). Commande des processus, collection Systèmes automatiques, tome 2, Chap. 2, Paris:Editions Ellipses , ISBN 2-7298-9720-8.

Hautier, J.P., & Caron, J.P. (1999). Convertisseurs statiques: Méthodologie causale de modélisation et de commande,Paris:Editions Technip, ISBN 2-7108-0745-9.

Hautier, J.P., & Faucher, J. (1996). Le graphe Informationnel Causal, outil de modélisation et de synthèse des commandes des processus électromécaniques, Bulletin de l'Union des Physiciens, no 785, cahier spécial de l'Enseignement Supérieur, pp. 167–189.

Louis, J.P., Multon, B., Bonnasieux, Y., & Lavabre, M. (1999). Commande des machines à courant continu à vitesse variable, Technique de l'ingénieur, traité de l'ingénieur, ref. D 3617, pp. 6–14, Paris.

Louis, J.P. (sous la direction de) (2004). Modèles pour la commande des actionneurs électriques, traité EGEM, Série Génie Electrique, Chapitre 1, Hermès- Lavoisier, ISBN 2-7462-0917-9.

Luenberger, D. G. (1966). Observers for multivariable systems. *IEEE Transactions on Automatic Control, 11*(2), 190–197.

Pietrzak-David, M., & de Fornel, B. (1994). Observation d'état déterministes et stochastiques dans la commande vectorielle d'un variateur asynchrone, Actes de la Journées d'Etudes SEE sur les Méthodes de l'Automatique appliquées à l'Electrotechnique, Lille, April 7th, 1994.

Vas, P. (1990). Vector control of AC machines. London:Oxford Science Publications, Clarendron Press, ISBN 0-19-859370-8.

Verghese, G., & Sanders, S. (1988). Observers for flux estimation in induction machine. *IEEE Transaction on Industrial Electronics, 35*(1), 85–94.

Von Westerholt, E.G. (1994). Commande non linéaire d'une machine asynchrone. Ph D Thesus report from Institut National Polytechnique de Toulouse.

Chapter 4
Theoretical Study of the Parametric Sensitivity

4.1 Position of the Problem

So as to avoid the measurement of the flux, the previous chapter has shown that it is possible to estimate or to impose the flux indirectly by using models. This great dependence on a model results in significant sensitivity problems of these controls to the uncertainties on this model. These uncertainties are due to the variations in the stator and rotor resistances according to the temperature and to the skin effect. They are also due to the variations in the inductances according to the magnetic saturation.

The rotor resistance is the most difficult parameter to identify with precision, especially for squirrel-cage machines, while it generally plays a great part in the vector control. This parameter may vary by 100% with the temperature. The stator resistance suffers a priori lower variations due to the temperature because the stator is generally close to the outer shell of the machine. However, the skin effect is higher for the stator conductors, since the stator current frequency may vary from 0 to 50 Hz when running normally and may reach higher values in a field weakening zone when the motor speed is higher than the rated speed. However, the skin effect is only significant when the wire radius is higher than the penetration depth (which is about 9 mm for the copper at 50 Hz and about 6.5 mm at 100 Hz). The rotor resistance is not affected by this phenomenon at all, as the rotor frequency is only a fraction of the stator currents' frequency, proportional to the slip. In general, the conduction resistance of the semiconductors that compose the inverter must be added to the stator resistance.

For instance, this resistance can reach some tenths of ohms for MOSFET transistors.

The magnetic saturation may result in variations of about 10–20% on the inductances. The variations of the ohmic and inductive parameters are coupled. Indeed, the errors on the values of stator and rotor resistances result in errors on the value of the magnetic flux into the machine, and thus change the value of the magnetization inductance.

B. Robyns et al., *Vector Control of Induction Machines*, Power Systems,
DOI: 10.1007/978-0-85729-901-7_4, © Springer-Verlag London 2012

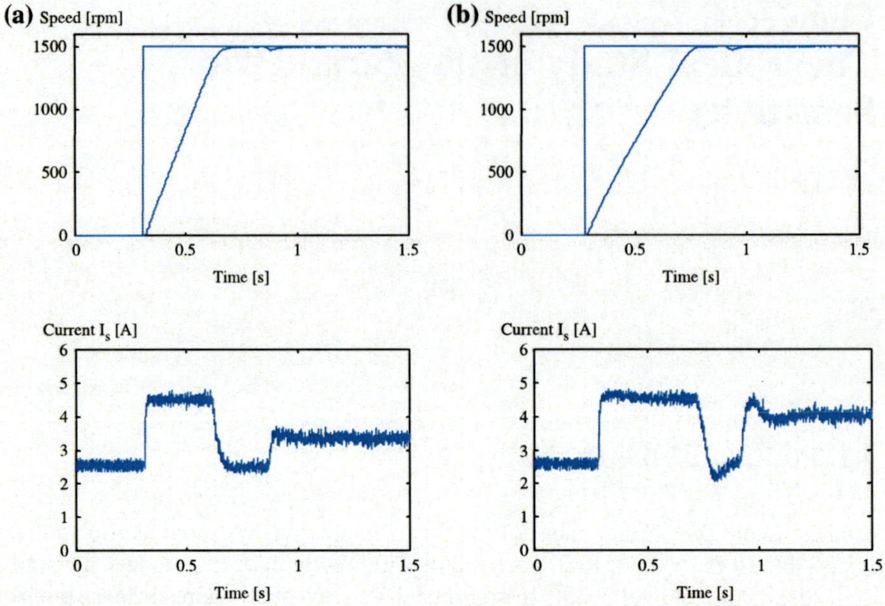

Fig. 4.1 Experimental response to a ramp function followed by a resisting torque step. **a** Optimized parameters. **b** With a 100% error on the estimated rotor resistance

Figure 4.1 shows the degradation of the performance of a standard vector control (Fig. 3.28), due to a significant error on the estimation of the rotor resistance. This figure shows the experimental response and the speed reference, as well as the response of the stator current root-mean-square value to a ramp signal from 0 to 1,500 rpm followed by a torque step of almost equal value to the rated torque (the characteristics of the tested induction machine are presented at the end of this chapter). The test shown in Fig. 4.1a has been realized with optimized parameters, whereas in the test of Fig. 4.1b, a 100% error on the estimated rotor resistance has been introduced.

The system response in Fig. 4.1b is significantly not as good as the system response in Fig. 4.1a because:

- the consumed stator current, when the motor is loaded, is increased by more than 20% to develop the same torque;
- the speed's response time is also increased because—for a similar consumed current during the acceleration phase—the developed torque does not reach its maximum value.

These degradations are due to the fact that the electromagnetic torque into the machine is no longer optimally created. It means that the current and the flux that generate this torque are no longer in quadrature. Moreover, the flux amplitude in general does not reach the value expected in the presence of parametric errors.

Thus, the uncertainties on the parameters result in errors on the amplitude and on the flux orientation into the machine with the following consequences:

- the response of the controlled quantities (speed, torque, flux, current) is damaged;
- the system may become unstable when the orientation error becomes too significant;
- an additional stator current is consumed to develop a given torque, which reduces the system efficiency and may need an oversizing of the inverter supplying the machine.

4.2 Theoretical Study of Parametric Sensitivity

4.2.1 Errors on the Magnetic Flux Control

4.2.1.1 Concept and Hypotheses

In order to analytically study the parametric sensitivity of the several control strategies while limiting the complexity of the equations to be manipulated, the theoretical analysis of sensitivity will be based on the equations of the induction machine in steady state. In spite of this steady-state hypothesis, instabilities can be predicted (when the orientation error becomes too significant). In addition, the conclusions of the theoretical study can generally be applied to transient-state running.

Figure 4.2 shows the different flux vectors that interfere in the studied system, in a reference frame related to the rotating field. The following vectors can be distinguished:

- the actual flux into the machine: ϕ;
- the estimated flux into the control: $\tilde{\phi}$;
- the reference flux: ϕ_{ref};

In the total absence of uncertainty on the machine model, these three vectors merge and perfectly match the reference flux. This situation is theoretical, and these three vectors may differ in practice. Therefore, the actual flux does not exactly match the reference flux. The length of these two vectors may differ, which results in an **error on the flux amplitude** written as follows:

$$\left|\frac{\phi}{\phi_{\text{ref}}}\right| = \frac{\sqrt{\phi_d^2 + \phi_q^2}}{\phi_{\text{ref}}} \tag{4.1}$$

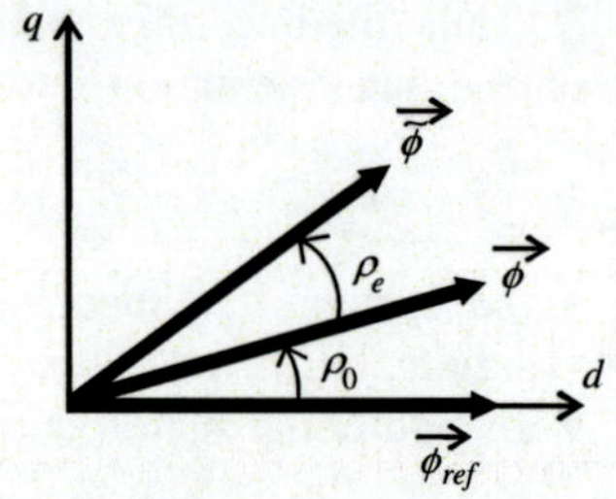

Fig. 4.2 Different flux vectors in a reference frame related to the rotating field

An angular offset may also appear between these two vectors, resulting in an error on the flux orientation, written as follows:

$$\rho_o = \text{arctg}\left(\frac{\phi_q}{\phi_d}\right) \tag{4.2}$$

With no error on the parameters, expressions (4.1–4.2) are reduced to $\phi = \phi_{\text{ref}}, \rho_0 = 0$. An angular offset may also exist between the estimated flux and the actual flux. This offset, known as estimation error of the flux phase, is written as ρ_e in Fig. 4.2.

In the case of controls with a priori, axis component q of the estimated flux is supposed null. The estimated flux vector, as well as the reference flux, are then oriented along axis d. The estimation error of the flux phase is then equal to the orientation error $(\rho_e = -\rho_0)$.

In the case of controls without a priori, both errors ρ_e and ρ_0 may differ according to the reference frame in which the flux estimator or flux observer is developed, and according to the technique of flux orientation that is implemented.

An error on the orientation of the flux assumes specific importance, as the system may become unstable when this error becomes more important (i.e. when it tends to $\pi/2$).

In some cases, it is interesting to express the errors on the flux amplitude and on the flux orientation in terms of sensitivity. The sensitivity function S_a of the flux amplitude and the sensitivity S_o of the flux orientation to the variations of a parameter are written as follows:

$$S_a = \frac{\frac{\phi_{\text{ref}} - \phi}{\phi_{\text{ref}}}}{\frac{\Delta X}{X^*}} \; and \; S_o = \frac{\rho_o}{\frac{\Delta X}{X^*}} \tag{4.3}$$

X is the studied parameter. $\Delta X = X^* - X$ in which the exponent (*) represents an estimated parameter.

It is difficult to experimentally measure the errors on the flux amplitude and on the flux orientation. However, these errors on the flux result in variations of the consumed stator current that can be measured. These variations allow the relation between the experimentation and the theoretical predictions and therefore allow the validation of the theoretical predictions. The variation of the stator current is written as:

$$\Delta I_s = \frac{I_s - I_{si}}{I_{si}} \tag{4.4}$$

in which I_{si} is the root-mean-square value of the consumed ideal current with no error on the parameters, and I_s the root-mean-square value of the current that is really consumed.

The variation of the consumed stator current will also be used as an additional criterion of comparison (added to the errors on the flux amplitude and on the flux orientation) of different control strategies.

The estimation of the flux components is necessary in some controls when developing electromagnetic torque and mechanical speed estimators. The theoretical study of sensitivity can be applied to detect, the errors on the estimation of the electromagnetic torque and on the estimation of the mechanical speed, if necessary. So as to determine the errors on the flux amplitude and on the flux orientation, the following elements must be considered separately:

- the controls in which the flux component of axis q is supposed null with a priori;
- the controls with estimation of flux components d and q without a priori.

Later in this chapter, the approach that allows the general definition of the errors on the flux is first presented. This approach is then applied to a few rotor-flux-oriented vector control strategies.

4.2.1.2 Case of Controls with a Priori

When the estimated flux component of axis q is supposed null, and when reminding that $\phi_{qref} = 0$ and $\phi_{dref} = \phi_{ref}$, the following equations are obtained from the machine model under steady-state conditions associated to the control strategy:

$$\begin{aligned} C\,\phi_{ref} &= A_1\,\phi_d - B_1\,\phi_q \\ D\,\phi_{ref} &= B_2\,\phi_d + A_2\,\phi_q \end{aligned} \tag{4.5}$$

The expressions of coefficients A_1, A_2, B_1, B_2, C and D depend on the actual and estimated parameters of the machine, the mechanical speed and the slip angular frequency. They are defined according to the control strategy and to the type of controlled flux (rotor or stator flux). They are developed in some cases later in this book. When resolving the system composed of Eq. 4.5, ϕ_d and ϕ_q can be written according to ϕ_{ref} :

$$\phi_d = \frac{B_1\,D + A_2\,C}{A_1\,A_2 + B_1\,B_2}\,\phi_{ref} \text{ and } \phi_q = \frac{A_1\,D - B_2\,C}{A_1\,A_2 + B_1\,B_2}\,\phi_{ref} \tag{4.6}$$

From expression (4.6), the following errors can be defined:

- the error on the flux amplitude:

$$\left|\frac{\phi}{\phi_{\text{ref}}}\right| = \frac{\sqrt{\phi_d^2 + \phi_q^2}}{\phi_{\text{ref}}} = \sqrt{\frac{(B_1 D + A_2 C)^2 + (A_1 D - B_2 C)^2}{(A_1 A_2 + B_1 B_2)^2}} \tag{4.7}$$

- the error on the flux orientation:

$$\rho_o = \text{arctg}\left(\frac{\phi_q}{\phi_d}\right) = \text{arctg}\left(\frac{A_1 D - B_2 C}{A_2 C + B_1 D}\right) \tag{4.8}$$

Expressions (4.7, 4.8) depend on rotor pulsation ω_r. The electromagnetic torque must be expressed according to coefficients A_1, A_2, B_1, B_2, C and D so as to determine the value of this pulsation in the presence of errors on the parameters. Whatever the torque expression used (Table 3.3), it can be written as follows:

$$T = pf(\omega_r)\left(\phi_d^2 + \phi_q^2\right) \tag{4.9}$$

From expressions (4.7), it then results that:

$$T = pf(\omega_r)\frac{(B_1 D + A_2 C)^2 + (A_1 D - B_2 C)^2}{(A_1 A_2 + B_1 B_2)^2}\phi_{\text{REF}}^2 \tag{4.10}$$

Expression $f(\omega_r)$ represents a function of the rotor pulsation whose shape depends on the type of the controlled flux (rotor or stator flux). So as to define this pulsation, Eq. 4.10 must be solved, in general numerically, because of the strong nonlinearity of this equation.

4.2.1.3 Case of Controls Without a Priori

In the case of controls without a priori, both components of the flux are assessed. It is then easier to deal with the complex equations of the machine with the estimator or the observer. The flux, current and voltage vectors are then written as follows:

$$\underline{\phi} = \phi_d + j\,\phi_q \quad \underline{\tilde{\phi}} = \tilde{\phi}_d + j\,\tilde{\phi}_q \tag{4.11}$$

$$\underline{I}_s = i_{\text{sd}} + j\,i_{\text{sq}} \quad \underline{V}_s = v_{\text{sd}} + j\,v_{\text{sq}} \tag{4.12}$$

The steady-state conditions can be taken into consideration—regardless of the referential in which the observer is developed, as quantities have a sinusoidal shape when they are not constant—in substituting the derivatives with their complex form:

$$\underline{\dot{X}} = j(\omega_s - \omega_p)\underline{X} \tag{4.13}$$

Pulsation ω_p has already been defined in Sect. 3.4.2. The equations of the flux observer can then be written as follows:

$$\underline{A}_o\,\underline{\tilde{\phi}} + \underline{B}_o\,\underline{I}_s + \underline{C}_o\,\underline{V}_s = 0$$

with

$$\underline{A}_o = A_{o1} + jA_{o2}, \underline{B}_o = B_{o1} + jB_{o2} \text{ and } \underline{C}_o = C_{o1} + jC_{o2} \tag{4.14}$$

Complex expressions $\underline{A}_o$, $\underline{B}_o$ and $\underline{C}_o$ depend on the type of estimator or observer studied. Later in this chapter, these expressions will be developed for a reduced-order rotor-flux observer (Eqs. 4.50 and 4.51).

To determine the errors on the estimated flux, stator currents and voltages in (4.14) have to be eliminated—expressing them according to the real flux by the equations of the machine—so as to obtain the following complex expressions:

- for the stator current:

$$\underline{I}_s = \underline{G}\,\underline{\phi} \text{ and } \underline{G} = G_1 + jG_2 \tag{4.15}$$

- for the stator voltage:

$$\underline{V}_s = \underline{H}\,\underline{\phi} \text{ and } \underline{H} = H_1 + jH_2 \tag{4.16}$$

The expressions of coefficients G_1, G_2, H_1 and H_2 depend on the kind of flux represented by ϕ (*rotor* or *stator* flux). When introducing (4.15) and (4.16) in (4.14), the result is:

$$\underline{A}_o\,\underline{\tilde{\phi}} + (\underline{B}_o\,\underline{G} + \underline{C}_o\,\underline{H})\;\underline{\phi} = 0 \tag{4.17}$$

A relation between the estimated flux and the real flux is deduced from expression (4.17):

$$\underline{\tilde{\phi}} = (q_1 + jq_2)\;\underline{\phi} \tag{4.18}$$

with

$$q_1 = \frac{-A_{o1}\,Z_1 - A_{o2}\,Z_2}{A_{o1}^2 + A_{o2}^2} \text{ and } q_2 = \frac{A_{o2}\,Z_1 - A_{o1}\,Z_2}{A_{o1}^2 + A_{o2}^2} \tag{4.19}$$

where

$$Z_1 = B_{o1}\,G_1 - B_{o2}\,G_2 + C_{o1}\,H_1 - C_{o2}\,H_2$$

and

$$Z_2 = B_{o2}\,G_1 + B_{o1}\,G_2 + C_{o2}\,H_1 + C_{o1}\,H_2$$

The expressions of coefficients q_1 and q_2 depend on the observer studied. These parameters depend on estimated and real parameters, on the mechanical speed and on the rotor pulsation. When there is no error on the parameters, then $q_1 = 1$ and $q_2 = 0$.

The **error on the estimation of the flux amplitude** is deduced from (4.18):

$$\left|\frac{\phi}{\tilde{\phi}}\right| = \frac{1}{\sqrt{q_1^2 + q_2^2}} \tag{4.20}$$

and the **error on the estimation of the flux phase:**

$$\rho_e = \text{arctg}\left(-\frac{q_2}{q_1}\right) \tag{4.21}$$

When the estimated flux is controlled by a controller with integral action, the result under steady-state conditions is $|\tilde{\phi}| = \phi_{\text{ref}}$. The **error on the flux amplitude** is then directly obtained from (4.20):

$$\left|\frac{\phi}{\phi_{\text{ref}}}\right| = \frac{1}{\sqrt{q_1^2 + q_2^2}} \tag{4.22}$$

The **orientation error of flux ρ_o** (Fig. 4.2) depends on the orientation technique used for the flux, which is the function of the reference frame in which the observer is developed.

Expressions (4.21) and (4.22) depend on the slip pulsation ω_r. To determine the value of this pulsation when errors are present on parameters, electromagnetic torque T has to be expressed in relation to coefficients q_1 and q_2. When an integral action is present in the flux controller, the electromagnetic torque developed by the machine is then [if (4.22) is taken into account]:

$$T = pf(\omega_r)|\phi|^2 = pf(\omega_r)\frac{\phi_{\text{ref}}^2}{q_1^2 + q_2^2} \tag{4.23}$$

In (4.23), $f(\omega_r)$ stands for a function of slip pulsation the shape of which depends on the type of controlled flux (rotor or stator flux). Equation 4.23 has to be solved to determine this slip pulsation. As for the case of control without a priori, this solution will generally be numerically obtained because the equation is highly nonlinear.

4.2.2 Application to Control with a Priori on the Rotor Flux

4.2.2.1 Influence of the Current Control Strategy

The control strategy studied in this section is an indirect vector control where the rotor flux is explicitly oriented (Fig. 3.27). For the calculation of the estimated e.m.f. (Fig. 4.3), the decoupling works on the references of the currents, not on their measurements.

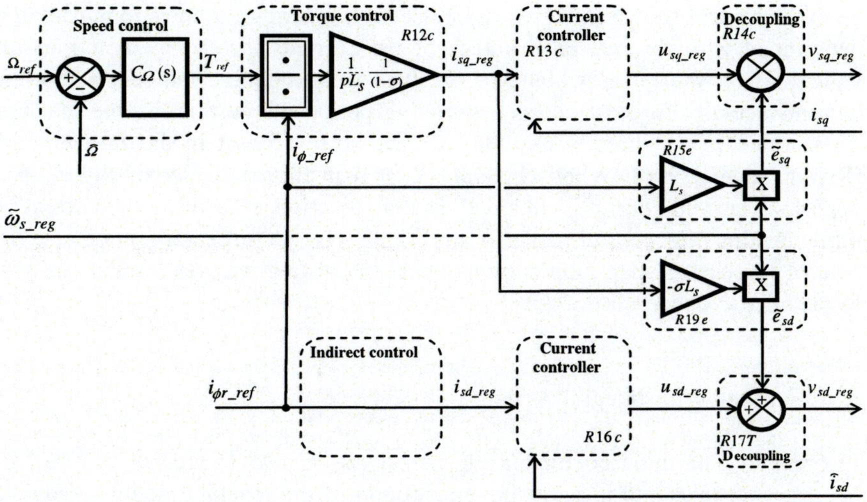

Fig. 4.3 Indirect control in using correcting variables for compensations

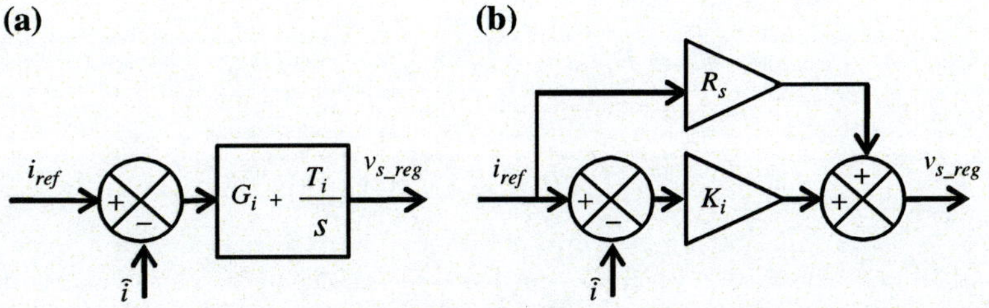

Fig. 4.4 Current PI controllers (**a**), combining a proportional feedback and a feedforward (**b**)

Furthermore, the steady-state conditions are supposed to be set up. Consequently, the estimation of the quadrature component of the stator e.m.f. is easier (Eq. 3.62).

$$R15e \rightarrow \tilde{e}_{\mathrm{sq}}(s) = L_s^* \hat{i}_{\mathrm{sd}}(s)\, \tilde{\omega}_{s\_\mathrm{reg}}(s) = L_s^* \hat{i}_{\phi r\_ref}(s)\; \tilde{\omega}_{s\_\mathrm{reg}}(s) \tag{4.24}$$

Various sorts of controllers will be studied to provide the control of currents i_{sd} and i_{sq}. The following element will be emphasized: the influence of the controller choice on the sensitivity of the flux control to parameter uncertainties.

The first controller studied is a standard proportional-integral (PI) controller (Fig. 4.4a). Nevertheless, a less standard controller offers interesting features. This is a controller which combines a proportional feedback with a feed forward action (also called PFF for Proportional Feed Forward) shown in Fig. 4.4b. From the controller of Fig. 4.4b, the standard proportional controller is obtained when $R_s = 0$.

In the control shown in Fig. 4.3, the decoupling terms and the slip pulsation are determined with the reference values of the currents instead of their measured values. Then, the terms which are obtained are predictive (since the reference currents actually forecast on each sampling period the currents in the machine). Their value is not directly modified by the noise present in the measurements [Robyns et al. 1996]. When choosing $K_i = 0$ in the controller of Fig. 4.4b, an open-loop current control is obtained. In that situation, measuring the currents for implementing the control strategy of Fig. 4.3 is not necessary anymore. This control strategy of open-loop currents offers interesting features toward sensitivity to parametric uncertainties.

A. Controller With Integral Action

In order to determine coefficients A_1, A_2, B_1, B_2, C and D present in (4.6), it is necessary to take into account the equations—of the machine under steady-state conditions—which are expressed in relation to the rotor-flux components. From Eq. 2.117, the following equations in steady-state condition are deduced $\left(v_{\mathrm{rd}} = v_{\mathrm{rq}} = 0\right)$:

$$v_{\mathrm{sd}} = R_s\, i_{\mathrm{sd}} - \sigma\, L_s\, \omega_s\, i_{\mathrm{sq}} - \frac{M}{L_r} \omega_s\, \phi_{\mathrm{rq}} \tag{4.25.a}$$

$$v_{\mathrm{sq}} = \sigma\, L_s\, \omega_s\, i_{\mathrm{sd}} + R_s\, i_{\mathrm{sq}} + \frac{M}{L_r} \omega_s\, \phi_{\mathrm{rd}} \tag{4.25.b}$$

$$i_{\mathrm{sd}} = \frac{1}{M} \phi_{\mathrm{rd}} - \frac{L_r}{M\, R_r} \omega_r\, \phi_{rq} \tag{4.25.c}$$

$$i_{\mathrm{sq}} = \frac{L_r}{M\, R_r} \omega_r\, \phi_{\mathrm{rd}} + \frac{1}{M} \phi_{\mathrm{rq}} \tag{4.25.d}$$

It is important to mention again that under steady-state conditions, the rotor flux is oriented $\left(\phi_{\mathrm{rd}} = \phi_{\mathrm{ref}}, \phi_{\mathrm{rq}} = 0\right)$. When currents i_{sd} and i_{sq} are controlled with a controller with integral action, the result under steady-state conditions is:

$$\begin{aligned} i_{\mathrm{sd}} &= i_{\mathrm{sd\_ref}} = \frac{1}{M^*} \phi_{r\_\mathrm{ref}} \\ i_{\mathrm{sq}} &= i_{\mathrm{sq\_ref}} = \frac{1}{M^*} \frac{L_r^*}{R_r^*} \omega_r\, \phi_{r\_\mathrm{ref}} \end{aligned} \tag{4.26}$$

When introducing (4.26) in the equations of flux (4.25.c and d), the result—by identification with (4.5) - is:

$$\begin{aligned} A_1 = A_2 = \frac{1}{M} \quad B_1 = B_2 = \frac{1}{M}\frac{L_r}{R_r}\omega_r \\ C = \frac{1}{M^*} = A_1^* \quad D = \frac{1}{M^*}\frac{L_r^*}{R_r^*}\omega_r = B_1^* \end{aligned} \tag{4.27}$$

It has to be emphasized that these expressions are independent from the mechanical speed. When introducing them in Eqs. 4.7, 4.8 and 4.10, it is then possible to determine errors on amplitude and on flux orientation caused by parametric errors. When introducing the expressions of the currents under steady-state conditions—Eq. 4.25.c and d—in the equation of torque (2.106), the result is:

$$T = p\frac{\omega_r}{R_r}\left(\phi_{rd}^2 + \phi_{rq}^2\right) \tag{4.28}$$

By identification with general expression (4.9), the result is:

$$f(\omega_r) = \frac{\omega_r}{R_r} \tag{4.29}$$

It is also possible to control alternating currents of the motor, instead of the Park components of these currents which change as continuous variables. If alternating currents are controlled by PI controller (either numerical or analogical controllers) or by resonant controllers (Hautier et al. 1999), the parametric sensitivity of the control under steady-state conditions is similar to the one obtained in regulating the Park components of these currents with PI controllers.

B. Controllers with Proportional and Feed Forward Actions

When currents are controlled with the controller shown in Fig. 4.4b—combining a proportional feedback and a feed forward—control voltages v_{sd} and v_{sq} then have the following values:

$$\begin{aligned} v_{sd\_reg} &= K_i(i_{sd\_ref} - \widehat{i}_{sd}) + R_s^* i_{sd\_ref} - \sigma^* L_s^* \omega_s i_{sq\_ref} \\ v_{sq\_reg} &= K_i(i_{sq\_ref} - \widehat{i}_{sq}) + R_s^* i_{sq\_ref} + \frac{L_s^*}{M^*}\omega_s \varphi_{r\_ref} \end{aligned} \tag{4.30}$$

where

$$\begin{aligned} i_{sd\_ref} &= \frac{1}{M^*}\phi_{ref} \\ i_{sq\_ref} &= \frac{1}{M^*}\frac{L_r^*}{R_r^*}\omega_r \phi_{r\_ref} \end{aligned} \tag{4.31}$$

When introducing (4.31) in (4.30), and (4.30) in stator equations (4.25), and when eliminating currents i_{sd} and i_{sq} with the equations of flux (4.25), it results in:

$$\begin{aligned} A_1 = A_2 &= \frac{1}{M}\left((R_s + K_i) - \sigma L_s \omega_s \frac{L_r}{R_r} \omega_r\right) \\ B_1 = B_2 &= \frac{1}{M}\left(\frac{L_r}{R_r} \omega_r (R_s + K_i) + L_s \omega_s\right) \\ C = A_1^* &= \frac{1}{M^*}\left((R_s^* + K_i) - \sigma L_s^* \omega_s \frac{L_r^*}{R_r^*} \omega_r\right) \\ D = B_1^* &= \frac{1}{M^*}\left(\frac{L_r^*}{R_r^*} \omega_r (R_s^* + K_i) + L_s^* \omega_s\right) \end{aligned} \quad (4.32)$$

The case where currents i_{sd} and i_{sq} are controlled with an ordinary proportional action controller—in imposing $R_s^* = 0$ in (4.30) and consequently in (4.32)—can easily be taken into account. Since there is still a static error in this controller and since this controller induces poor performance as far as the parametric sensitivity of the flux control is concerned, this controller will not be studied any longer here (Robyns et al. 1994). On the contrary, when imposing $K_i = 0$ in (4.30), the result is a control without current regulation whose relevant performance is mentioned later on in this chapter.

C. Introduction of Magnetic Saturation

Expressions (4.27) and (4.32) are calculated from a linear model which does not take magnetic saturation into account. Since errors on the estimated values of the stator and rotor resistances induce errors on the flux value in the machine and then change the value of the magnetization inductance, an ordinary model modeling the variations of parameter M is introduced into the sensitivity study. The normalized stator flux ϕ_{sn} (Sect. 2.5) is linked to the rotor flux by the following relation:

$$\phi_{sn} = \frac{\frac{L_s}{M}\sqrt{1 + \left(\sigma \frac{L_r}{R_r} \omega_r\right)^2}}{\frac{L_s^*}{M^*}\sqrt{1 + \left(\sigma^* \frac{L_r^*}{R_r^*} \omega_r\right)^2}} \sqrt{\frac{(B_1 D + A_2 C)^2 + (A_1 D - B_2 C)^2}{(A_1 A_2 + B_1 B_2)^2}} \frac{\phi_{r\_ref}}{\phi_{Nref}} \quad (4.33)$$

where ϕ_{Nref} is the rated value of the reference flux.

To obtain the error on the amplitude and flux orientation, a system composed of two Eqs. 4.10 and 2.137 with two unknowns has to be solved for every operating point determined by a value of speed and of electromagnetic torque T. The two unknowns are rotor pulsation ω_r and inductance M. As this system is highly linear, it has to be numerically solved. In this study, the mechanical speed is supposed to be correctly measured from a speed or a position sensor.

Figure 4.5 shows the influence of magnetic saturation on the sensitivity of amplitude and orientation in relation to the electromagnetic torque and to an error on rotor resistance R_r. The control under consideration is represented in Fig. 4.3, PI controllers being on the stator currents. The curves in solid lines were obtained without taking into account magnetic saturation whereas the curves in dashed lines were obtained taking into account magnetic saturation. Figure 4.5 points out that

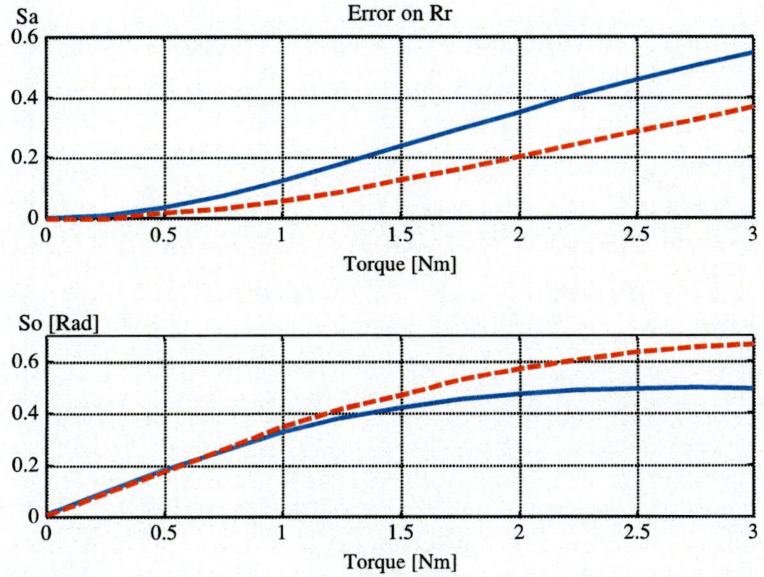

Fig. 4.5 Sensitivity of amplitude S_a and orientation S_o of the flux to errors on the rotor resistance. *Solid lines*: without saturation; *Dashed lines*: with saturation

magnetic saturation reduces the amplitude sensitivity but increases the flux orientation sensitivity, which can modify the system stability.

D. Variation of Consumed Stator Current

The variation of consumed stator current because of parametrical uncertainties is defined by Eq. 4.4. From the equations of flux (4.26), where $\phi_{rq} = 0$ and $\phi_{\mathrm{rd}} = \phi_{r\_\mathrm{ref}}$, the following expression is obtained for current I_{si} :

$$I_{\mathrm{si}} = \sqrt{i_{\mathrm{sdi}}^2 + i_{\mathrm{sqi}}^2} = \frac{\phi_{r\_\mathrm{ref}}}{M^*}\sqrt{1 + \left(\frac{\omega_{\mathrm{ri}} L_r^*}{R_r^*}\right)^2} \tag{4.34}$$

ω_{ri} is deduced from (4.10) and (4.28) if there is no parametric error:

$$\omega_{\mathrm{ri}} = \frac{T R_r^*}{p\phi_{r\_\mathrm{ref}}^2} \tag{4.35}$$

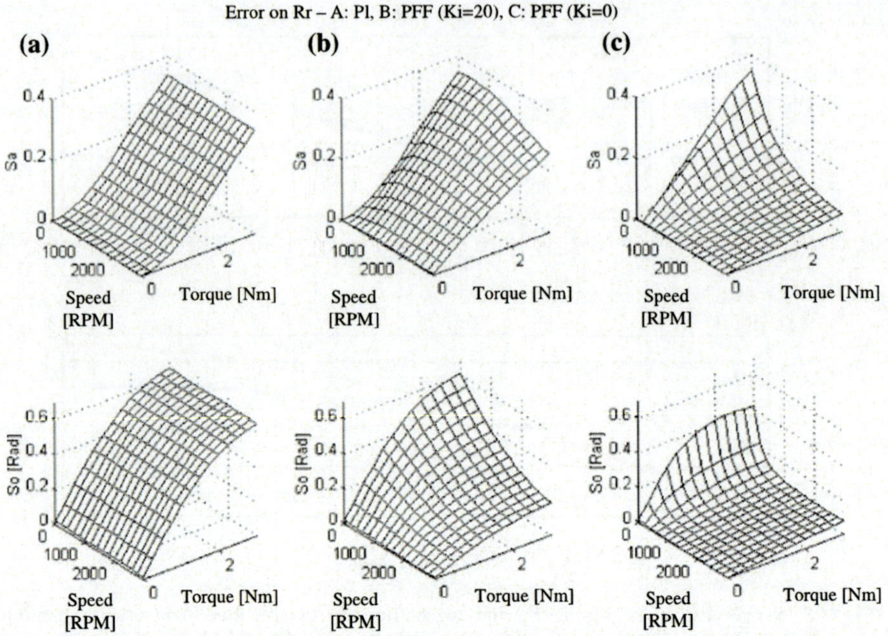

Fig. 4.6 Sensitivity of flux amplitude and orientation to uncertainties on R_r for various current control strategies. **a** PI. **b** PFF with $K_i = 20$. **c** PFF with $K_i = 0$

The expression of I_s is deduced from Eq. 4.25.c and 4.25.d:

$$I_s = \sqrt{i_{sd}^2 + i_{sq}^2} = \frac{\sqrt{\phi_{rd}^2 + \phi_{rq}^2}}{M}\sqrt{1 + \left(\frac{L_r}{R_r}\omega_r\right)^2} = \frac{\sqrt{\phi_{rd}^2 + \phi_{rq}^2}}{M}\sqrt{1 + (\tau_r\,\omega_r)^2} \tag{4.36}$$

When expression (4.7) is taken into account, I_s is:

$$I_s = \frac{\phi_{r\_ref}}{M}\sqrt{1 + (\tau_r\,\omega_r)^2}\sqrt{\frac{(A_2C + B_1D)^2 + (A_1D - B_2C)^2}{(A_1A_2 + B_1B_2)^2}} \tag{4.37}$$

E. Comparative Study of Various Solutions

Theoretical comparison:

Figures 4.6, 4.7 and 4.8 point out the sensitivity of flux amplitude and orientation in relation to errors on R_r R_s and M-respectively- and depending on the mechanical speed and on the electromagnetic torque. To obtain this, an error of 1% is introduced into the parameter which is under consideration. Curves A of each

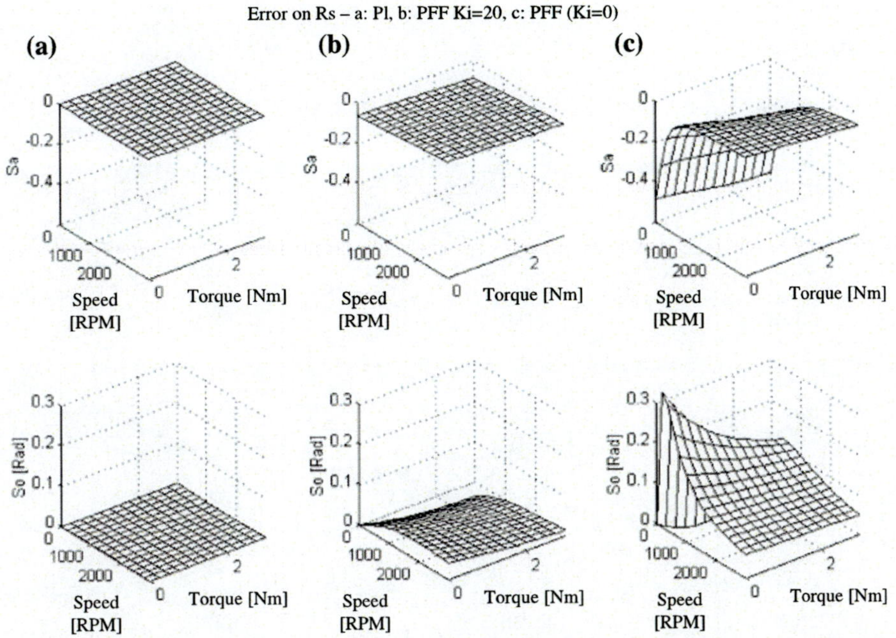

Fig. 4.7 Sensitivity of flux amplitude and orientation to uncertainties on Rs for various current control strategies. **a** PI. **b** PFF with $K_i = 20$. **c** PFF with $K_i = 0$

figure correspond to a control with PI controllers while curves B and C correspond to controls where currents are controlled by controllers combining a proportional feedback and a feedforward with $K_i = 20$ and $K_i = 0$, respectively.

Figures 4.6 and 4.8 show that errors on the flux are generally less important when currents are controlled by controllers combining a proportional feedback and a feedforward, rather than PI controllers.

However, in such a case, an error on the stator resistance has obviously no influence on the flux value under steady-state conditions. It should nonetheless be noted that the effect of an error on the value of R_s remains light and decreases quickly when speed grows. Besides, parameter R_s is one of the easiest to determine.

In the presence of an integral action, the gain of the controller does not affect the parametric sensitivity under steady-state conditions. However, if there is no integral action, K_i gain influences strongly the real flux value.

Figures 4.6–4.8 show that a high value of K_i gain offers a clear advantage only in presence of an error on R_s. Figure 4.6 in particular shows that the lowest sensitivity to uncertainties on R_r is obtained with the control without feedback of the measured current ($K_i = 0$).

Figure 4.9 emphasizes the additional consumption of stator current in the presence of a 100% error on the stator resistance $\left(R_r = 2R_r^*\right)$, in function of K_i

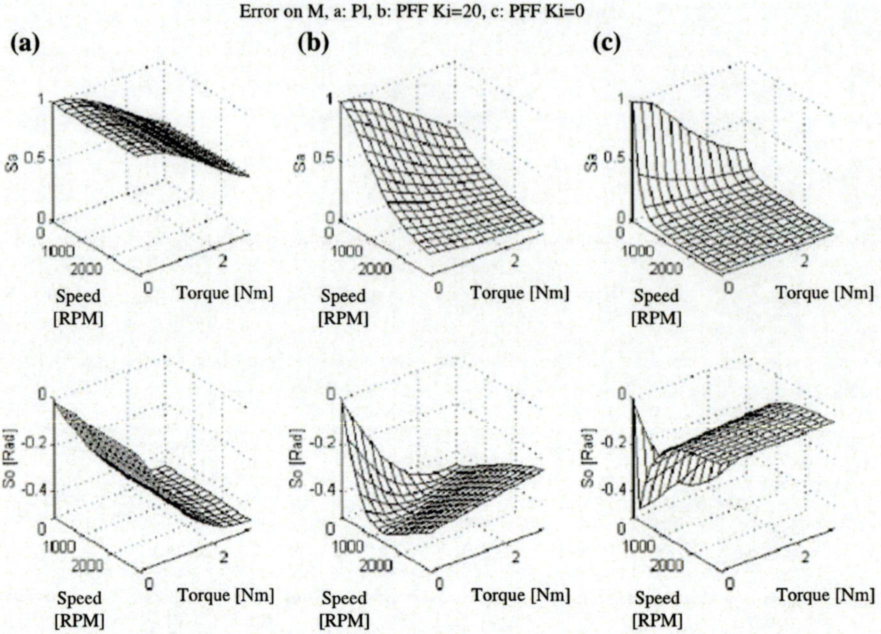

Fig. 4.8 Sensitivity of flux amplitude and orientation to uncertainties on M for various current control strategies (magnetic saturation is not taken into account in this case). **a** PI. **b** PFF with $K_i = 20$. **c** PFF with $K_i = 0$

gain for the following mechanical operating point: 1,500 rpm and 2.3 Nm. When currents are controlled by PI controllers, curves in continuous lines are obtained when magnetic saturation is taken into account and in discontinuous lines when magnetic saturation is not taken into account. When currents are controlled by PFF controllers, curves written * are obtained when magnetic saturation is taken into account and curves written + are obtained when magnetic saturation is not taken into account.

Figure 4.9 shows that magnetic saturation highly influences flux sensitivity and stator current variation. Ignoring saturation would lead to false conclusions. The figure shows–from the studied operating point—that the PFF controller with a gain inferior to 10 is less sensitive to errors on R_r than the PI controller.

When $K_i = 0$ control without currents feedback - the overcurrent consumption reduces from more than 50% in comparison to the PI controllers control. That is an interesting point concerning losses and the sizing of the inverter.

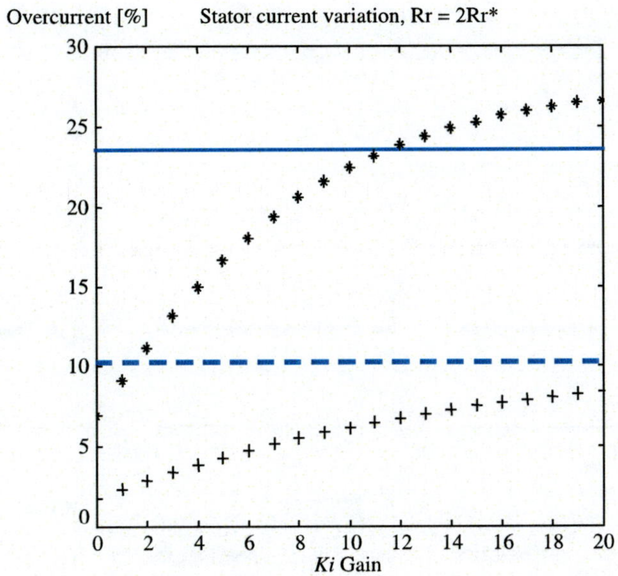

Fig. 4.9 Increase in the consumption of stator current in function of K_i $\left(R_r = 2R_r^*\right)$ Curve in *discontinuous line*: PI without magnetic saturation. Curve in *continuous line*: PI with magnetic saturation. Curve noted +: PFF without magnetic saturation. Curve noted *: PFF with magnetic saturation

E. Experimental comparison:

Tests were carried out with a test bench equipped with the TMS320C31 signal processor. The inverter associated to the induction motor of 750 W is equipped with MOSFET transistor, working at a commutation frequency of 30 kHz.

The other characteristics are presented at the end of this chapter. Figures 4.10 and 4.11 show that the system's response to a reference speed step from 0 to 1,500 rpm followed by a load torque step T_{load} from 0 to 2.3 Nm.

In the same way as for Figs. 4.6–4.8, curves A correspond in each figure to a control with PI current controller. While curves B and C correspond to controls in which currents control is made by controllers that associate a proportional feedback and a feed forward action (PFF), respectively with $K_i = 20$ and $K_i = 0$.

For each case, Figs. 4.10 and 4.11 show the mechanical speed reference values and measured values, as well as the measured values of the currents i_{sd} and i_{sq}.

Figure 4.10 shows the system's response when there is no error in parameters—it refers naturally to a verified hypothesis in this case; for example, when comparing tests to simulations or comparing the measured value of the consumption of the stator current to the theoretically predicted value.

Responses in speed are very similar in all three cases. Robustness of the different control strategies to the load variations is very satisfactory, even when there is no current control (case C). Figure 4.10c shows that, despite the sensitivity to

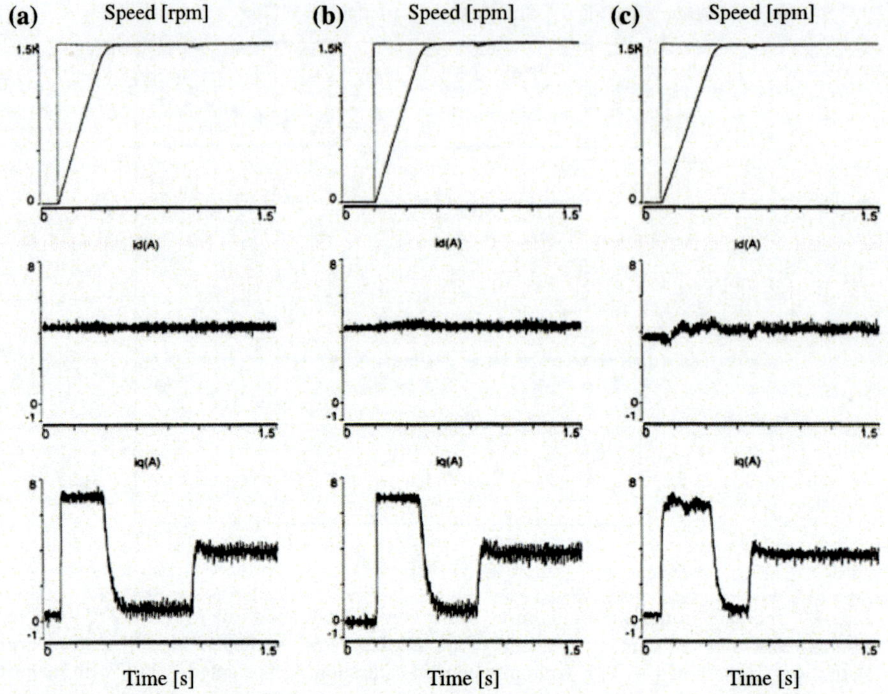

Fig. 4.10 System's response to a reference speed step from 0 to 1,500 rpm followed by a load torque step for different strategies of current control. Case **a** PI, Case **b** PFF with $K_i = 20$. Case **c** PFF with $K_i = 0$

Table 4.1 Increase in the consumption of stator current due to poor flux control, which is created by a 100% error on R_r (mechanical operating point: 1,500 rpm and 2.3 Nm)

Stator current variation			
Current controller	Experimental values [%]	Theoretical values with saturation [%]	Theoretical values without saturation [%]
PI	25	24.3	10.8
PFF $K_i = 20$	30	27.2	8.9
PFF $K_i = 10$	25	23.2	6.7
PFF $K_i = 5$	21	17	4.5
PFF $K_i = 3$	15.5	13.8	3.5
PFF K_i=2	13.5	11.4	3
PFF K_i=1	10	9.3	2.8
PFF K_i=0	8.6	7.4	1.9

uncertainties on R_s, current value i_{sd} is very close to its reference value when speed increases.

Figure 4.11 shows the system's response when a 100 % error is introduced on the rotor resistance $\left(R_r = 2R_r^*\right)$. The speed response slows down in the three cases, though curves remain very close to each other. However, the responses of

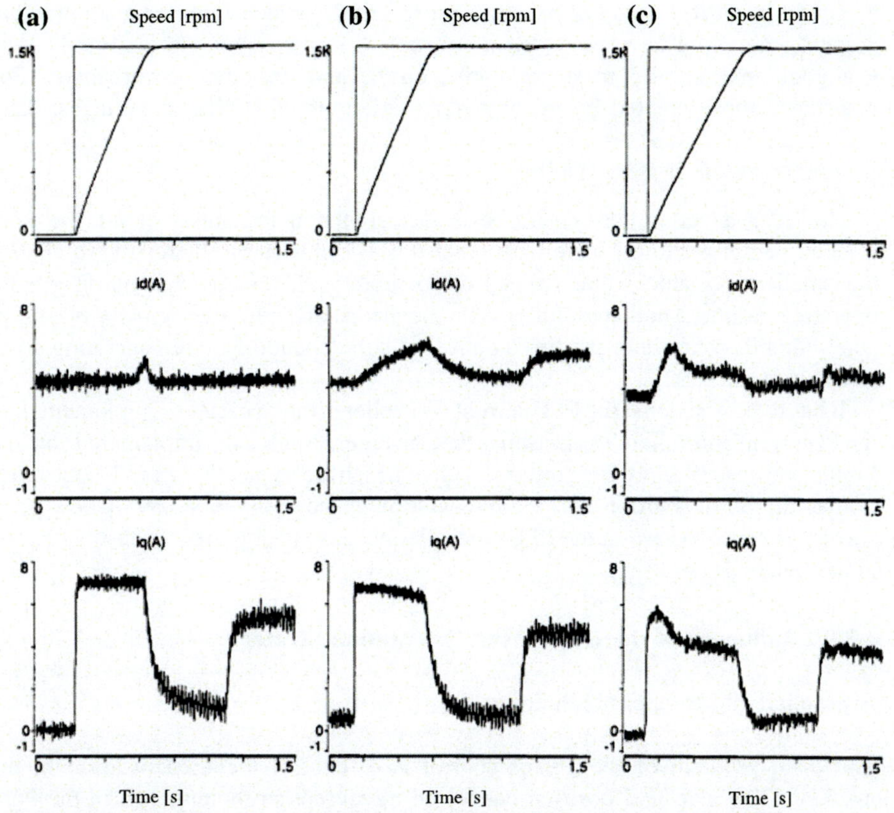

Fig. 4.11 System's response to a reference speed step from 0 to 1,500 rpm followed by a load torque step with a 100% error introduced on R_r Currents regulated by: **a** PI, **b** PFF with $K_i = 20$, and **c** PFF with $K_i = 0$

the current—that is to say of the torque—are very different in each case because of the different parametric sensitivities.

A comparison of the different tests can be done by measuring the consumption of stator current charging under steady-state conditions. Measured values are listed in Table 4.1 for different K_i values. They are also compared to the theoretical values. Experimental results listed in the table confirm the comments deduced from Fig. 4.9.

Tests carried out at a low speed and while braking are presented in the bibliographic reference (Robyns et al. 1999). The results that are presented in this section confirm the great influence current control strategy has on the parametric sensitivity of indirect vector controls. The current controllers—including an integral action—are not necessarily the best solution.

The study presented in this section shows that when currents are controlled by controllers combining a proportional feedback and a feed forward action (PFF), the K_i gain value should be a compromise between:

- a good insensitivity to uncertainties on R_r and M, which imposes a small value for K_i;
- a good insensitivity to uncertainties on R_s and on other disturbances. For instance those created by the inverter (variations of DC input voltages, dead time, etc.) that imposes a great value for K_i, which should however be limited in order to ensure system stability.

However, uncertainties on rotor resistance are the main issue since it is the most difficult parameter to estimate with precision. It can then be interesting to choose the smallest possible value for K_i, or to modify K_i's value depending on the operating point. Another solution can be deduced from the sensitivity study: combining PI controllers and PFF controllers still depending on the operating point [Robyns et al. 2002].

It has to be noted that the PFF current controller is much easier to implement than the PI current controller. The reason is that the latter requires the implementation of a limiter and an anti-windup algorithm, both of which are useless with a PFF controller.

The simplest solution for implementation is naturally obtained when K_i is chosen as equal to zero in the PFF controller because then current measures are no longer necessary.

4.2.2.2 Influence of the Rotor-Flux Orientation Strategy

A. Explicit and Implicit Technique

The main problem of the vector control is to control the field position in the machine. The real field position cannot be measured, so the angle θ_s in the Park transform is determined when the flux component of axis q is assumed to be null with the integration of the instantaneous stator frequency ω_s.

In Chap. 3 (Sect. 3.3.2.4), it has been proven that frequency ω_s could be determined by an explicit technique from the mechanical speed Ω and from an estimation of the slip frequency ω_r (estimation from Eq. 3.70):

$$\tilde{\omega}_{s1} = p\widehat{\Omega} + \tilde{\omega}_{r\_\text{reg}} = p\widehat{\Omega} + \frac{R_r^*}{L_r^*}\frac{\widehat{i_{sq}}}{i_{\phi r\_ref}} \tag{4.38}$$

The main inconvenience of this estimation is that it depends on the rotor resistance. It also depends on the inductance, but not on the stator resistance.

Stator frequency ω_s can also be determined from the stator equation of axis q of the machine model. In Chap. 3 (Sect. 3.3.2.5, Eq. 3.74), it has been shown that with an implicit technique a direct estimation of frequency ω_s is obtained:

$$\tilde{\omega}_{s2} = \frac{v_{sq\_reg} - (R_s^* + \sigma^* L_s^* s)\,\widehat{i_{\text{sq}}}}{L_s^* i_{\phi r\_\text{ref}}} \tag{4.39}$$

The main advantage of this expression is that it does not depend on rotor resistance. However, it depends on the stator resistance, on the stator inductance and on the stator current derivative of axis q. The derivative can generally be ignored.

B. Sensitivity Study

Theoretical expressions:

A strategy of indirect vector control in which currents are regulated by PI controllers is studied. When ω_s is determined by an explicit technique (Eq. 4.38), the diagram of the control structure in Fig. 4.3 is found again. Coefficients A_1, A_2, B_1, B_2, C and D that allow calculating errors on flux amplitude and orientation are then determined by (4.27).

When ω_s is determined by implicit technique (Eq. 4.39) coefficients A_1, B_1 and C are still determined by (4.27). On the contrary, coefficients A_2, B_2 and D are different. By expressing current i_{sq} from (4.39), introducing it into (4.25.d) and suppressing v_{sq}, i_{sd} and i_{sq} with (4.25 a,b,c), by identification with (4.5), we have:

$$\begin{aligned} D'' &= \frac{1}{M^*}\frac{L_s^*}{R_s^*}\omega_s \\ B_2'' &= \frac{1}{M}\frac{\omega_s}{R_s^*}\left(\sigma L_s + \frac{M^2}{L_r}\right) + \frac{1}{M}\left(\frac{R_s}{R_s^*} - 1\right)\frac{L_r}{R_r}\omega_r \\ A_2'' &= \frac{1}{M}\left[\left(\frac{R_s}{R_s^*} - 1\right) - \frac{\sigma L_s \omega_s}{R_s{}^*}\frac{L_r \omega_r}{R_r}\right] \end{aligned} \tag{4.40}$$

Theoretical constraint:

When ω_s is determined by implicit technique (Eq. 4.39), from the theoretical study is deduced the following expression for the flux orientation error:

$$\frac{\phi_{\mathrm{rq}}}{\phi_{\mathrm{rd}}} = \frac{\omega_s\left(1 - \frac{L_s}{L_s^*}\right) - \omega_r \frac{L_r}{R_r L_s^*}(R_s - R_s^*)}{\omega_s \omega_r \frac{L_r}{R_r}\left(1 - \frac{L_s}{L_s^*}\right) + \frac{1}{L_s^*}(R_s - R_s^*)} \tag{4.41}$$

If $\omega_r = 0$, the flux orientation error (4.2) becomes:

$$\rho_0 = \operatorname{arctg}\left(\frac{\phi_{rq}}{\phi_{rd}}\right) = \operatorname{arctg}\left(p\Omega \frac{L_s^* - L_s}{R_s - R_s^*}\right) \tag{4.42}$$

The expression (4.42) shows that for $\omega_r = 0$ with an error on inductance L_s, the orientation error (ρ_0) approaches $\pi/2$, which destabilizes the system. This result indicates then that the implicit technique should be used carefully.

4.2.2.3 Influence of the Estimation Strategy of the Rotor Flux

A. Estimation of the Flux

In the two previous sections it was imposed that rotor flux was in open loop from a simple model connecting the stator current component of axis d to the flux. In this section, estimators of the rotor-flux component of axis d are presented so that the component of axis q is assumed to be null. A PI controller that controls flux estimation is then introduced into the control. A direct vector control is thus obtained.

The simplest estimation of the magnetizing current that corresponds to the rotor flux was introduced in Chap. 3 (expression 3.63):

$$\tilde{i}_{\phi 1} = \frac{1}{1 + \tau_r^* s}\hat{i}_{sd} \tag{4.43}$$

In this expression, the magnetizing current is deduced from the measured current i_{sd}. It is also possible to estimate the magnetizing current from voltage v_{sd_reg} (Eq. 3.66):

$$\tilde{i}_{\phi 2} = \frac{1}{R_s^*}\frac{1}{\sigma^* \tau_s^* \tau_r^* s + (\tau_s^* + \tau_r^*)s + 1} u_{sd\_reg} \tag{4.44}$$

with

$$u_{sd\_reg} = v_{sd\_reg} + \sigma^* L_s^* \omega_s \hat{i}_{sq}$$

B. Sensitivity Study

Theoretical expressions:

When the magnetizing current is estimated with expression (4.43) and when the flux corrector has an integrator, we deduce in steady state:

$$\frac{\phi_{r\_ref}}{M^*} = i_{\phi r\_ref} = \tilde{i}_{\phi 1} = i_{sd} \tag{4.45}$$

By introducing (4.45) into the expression (4.25.c), we have:

$$A_1' = \frac{1}{M}, \quad B_1' = \frac{1}{M}\frac{L_r}{R_r}\omega_r \text{ and } C' = \frac{1}{M^*} \tag{4.46}$$

These expressions are the same as expressions (4.27). When magnetizing current is determined by (4.45), we then have under steady-state conditions:

$$i_{\phi r\_ref} = \frac{\phi_{r\_ref}}{M^*} = \tilde{i}_{\phi 2} = \frac{u_{sd\_reg}}{R_s^*} = \frac{v_{sd\_reg} + \sigma^* L_s^* \omega_s i_{sq}}{R_s^*} \tag{4.47}$$

By replacing v_{sd_reg} in (4.47) by its expression (4.25.a) and replacing i_{sd} and i_{sq} in (4.25.a) with (4.25.c and d), we have:

$$\begin{aligned} A_1'' &= \frac{1}{M R_s^*}\left[R_s - \omega_r \tau_r (\sigma L_s \omega_s - \sigma^* L_s^* \omega_s)\right] \\ B_1'' &= \frac{1}{M R_s^*}\left[\omega_r \tau_r R_s + (L_s \omega_s - \sigma^* L_s^* \omega_s)\right] \\ C'' &= \frac{1}{M^*} \end{aligned} \tag{4.48}$$

The expressions of the coefficients A_2, B_2 and D are the same as the corresponding expressions (4.27) if the controller of the current component of axis q includes an integral action.

It has to be noted that to determine coefficients that intervene in the sensitivity study, the nature of the controller of the current component of axis d was not taken into account. This means that the nature of the controller does not influence the sensitivity of the control under steady-state conditions. Therefore, as far as sensitivity is concerned, an integral action in this controller would appear redundant in comparison to the integral action of the flux controller.

Theoretical constraint

In the vector control, an estimation of the pulsation ω_s (4.38) or (4.39) must be associated to each estimation for the amplitude of the magnetizing current (and also of the flux). Therefore, several combinations are possible. These different combinations present different properties concerning their parametric sensitivity.

The parametric sensitivity of the combination of the estimation (4.43) with the explicit technique (4.38) is the same as the one presented in Sect. 4.2.2.1 when currents are regulated by controllers with integral actions.

The parametric sensitivity of the combination of the estimation (4.43) with the implicit technique (4.39) presents the theoretical constraint (4.42).

If estimation (4.44) is combined with (4.39), the following expression for the flux orientation error can be deduced from the theoretical study when $\omega_r = 0$:

$$\rho_o = \text{arctg}\left(\frac{\phi_{\text{rq}}}{\phi_{\text{rd}}}\right) = \text{arctg}\left(\frac{p\widehat{\Omega}\left(\frac{R_s}{R_s^*} - \frac{L_s}{L_s^*}\right)}{\frac{1}{L_s^*}\left(R_s - R_s^*\right) + p^2\,\widehat{\Omega}^2\left(\frac{L_s - \sigma^* L_s^*}{R_s^*}\right)}\right) \tag{4.49}$$

This expression shows that when there is no error on R_s, but one on L_s, when $\omega_r = p\widehat{\Omega} = 0$, the orientation error approaches $\pi/2$, which destabilizes the system. Therefore this combination is to be used carefully.

Finally, the combination of (4.44) and (4.38) gives a sensitive control on every parameter R_r, R_s and on the inductances.

4.2.3 *Application to Controls Without a Priori*

4.2.3.1 Rotor-Flux Observer into the Stator Reference Frame

A. Error on the Rotor Flux Amplitude

The principle scheme of a direct vector control based on the estimation of the rotor flux, obtained through an observer with a reduced order developed in the stator frame of reference, is introduced in Fig. 3.33. There are various controllers including generally an integral action. The following sections will show that, as the flux controller has an integral action, the nature of the current controllers does not have any influence on the sensitivity to parametric uncertainties under steady-state conditions. It is the same for the presence or the absence of decoupling terms.

In order to determine the errors induced by the parametric uncertainties on the flux amplitude, when the latter is estimated by a reduced-order observer, the expressions (4.20) or (4.22) must be determined. Expression (4.22) is only valid if the flux controller includes an integral action. To determine these expressions, coefficients q_1 and q_2 have to be developed in the context of the reduced-order observer. These coefficients are determined by the expressions (4.19) in which the complex coefficients expressions A_0, B_0 and C_0, characterizing the considered observer, have to be introduced.

At steady state and under complex form, system (3.85) takes the following form:

$$\begin{aligned} j(\omega_s - \omega_p)\underline{\tilde{\phi}}_r = (a_{1k} + ja_{2k})\underline{\tilde{\phi}}_r + (b_{1k} + jb_{2k})\underline{I}_s + \cdots + (-k_1 \\ - jk_2)\left[\underline{V}_s - \sigma^* L_s^* j(\omega_s - \omega_p)\underline{I}_s\right] \end{aligned} \tag{4.50}$$

Both the identification of this equation with (4.14) and the consideration of the expressions (3.88–3.91), give:

$$\begin{aligned} a_{o1} = a_{1k} = -\frac{R_r^*}{L_r^*} - k_1\frac{M^* R_r^*}{L_r^{*2}} - k_2\frac{M^*}{L_r^*}p\,\Omega \\ a_{o2} = a_{2k} - (\omega_p - p\,\Omega) = -\omega_r + k_1\frac{M^*}{L_r^*}p\,\Omega - k_2\frac{M^* R_r^*}{L_r^{*2}} \\ b_{o1} = b_{1k} - k_2\,\sigma^*\,L_s^*(\omega_s - \omega_p) = M^*\frac{R_r^*}{L_r^*} + k_1\,R_{sr}^* - k_2\,\sigma^*\,L_s^*\omega_s \\ b_{o2} = b_{2k} + k_1\,\sigma^*\,L_s^*(\omega_s - \omega_p) = k_1\,\sigma^*\,L_s^*\omega_s + k_2 R_{\mathrm{sr}}^* \\ c_{o1} = -k_1 \\ c_{o2} = -k_2 \end{aligned} \tag{4.51}$$

These expressions do not depend on pulsation ω_p (neither do they depend on the reference frame in which the observer is developed) except through gains k_1 and k_2

When coefficients q_1 and q_2 are determined, the error on the flux amplitude is calculated due to expression (4.22) in the case (most frequent) of a flux controller including an integral action.

B. Error on the Rotor-Flux Orientation

The orientation error of the rotor flux is defined by (Eq. 4.8):

$$\rho_o = \mathrm{arctg}\left(\frac{\phi_{rq}}{\phi_{\mathrm{rd}}}\right) \tag{4.52}$$

This expression is determined from the flux components in a reference frame related to the rotating field, whereas the considered observer of this chapter estimates the flux components in a fixed reference frame related to the stator (Sect. 3.4.4). In this context:

$$\begin{aligned} \phi_{rd} &= \phi_{r\alpha}\cos(\theta_s) + \phi_{r\beta}\sin(\theta_s) \\ \phi_{rq} &= -\phi_{r\alpha}\sin(\theta_s) + \phi_{r\beta}\cos(\theta_s) \end{aligned} \tag{4.53}$$

Replacing $\cos(\theta_s)$ and $\sin(\theta_s)$ by their expressions (3.109) gives:

$$\frac{\phi_{rq}}{\phi_{rd}} = \frac{-\phi_{r\alpha}\,\tilde{\phi}_{r\beta} + \phi_{r\beta}\,\tilde{\phi}_{r\alpha}}{\phi_{r\alpha}\,\tilde{\phi}_{r\alpha} + \phi_{r\beta}\,\tilde{\phi}_{r\beta}} \tag{4.54}$$

From relation (4.18), the following equations are obtained between the estimated flux components and the real flux components:

$$\begin{aligned} \tilde{\phi}_{r\alpha} &= q_1\,\phi_{r\alpha} - q_2\,\phi_{r\beta} \\ \tilde{\phi}_{r\beta} &= q_2\,\phi_{r\alpha} + q_1\,\phi_{r\beta} \end{aligned} \tag{4.55}$$

Including these equations in ratio (4.54) gives:

$$\frac{\phi_{rq}}{\phi_{rd}} = \frac{-q_2\left(\phi_{r\alpha}^2 + \phi_{r\beta}^2\right)}{q_1\left(\phi_{r\alpha}^2 + \phi_{r\beta}^2\right)} = -\frac{q_2}{q_1} \tag{4.56}$$

It can thus be concluded that the flux orientation error is identical to the estimation error of phase (4.21) for the considered flux observer:

$$\rho_o = \operatorname{arctg}\left(\frac{\phi_{rq}}{\phi_{rd}}\right) = \operatorname{arctg}\left(-\frac{q_2}{q_1}\right) = \rho_e \tag{4.57}$$

4.2.3.2 Choice of the Observer's Gains

The study of the sensitivity introduced in the previous section enables to study the influence of the gains and observer's poles on the sensitivity to electrical parametric uncertainties.

In Sect. 3.4.3.3, it has been explained that gains could be chosen in order to fix the observer poles. Nonetheless, this calculation method often results in an observer that is very sensitive to parametric uncertainties. It can also result in an unstable observer (even when the poles have a real negative part) because of a wrong excess in the flux orientation. Figure 4.12 illustrates this problem by showing the evolution of amplitude and flux phase errors, estimated by the reduced-order observer in the frame of the stator, depending on the real part of the poles in front of a 100% error on the estimation of the rotor resistance $\left(R_r = 2\,R_r^*\right)$.

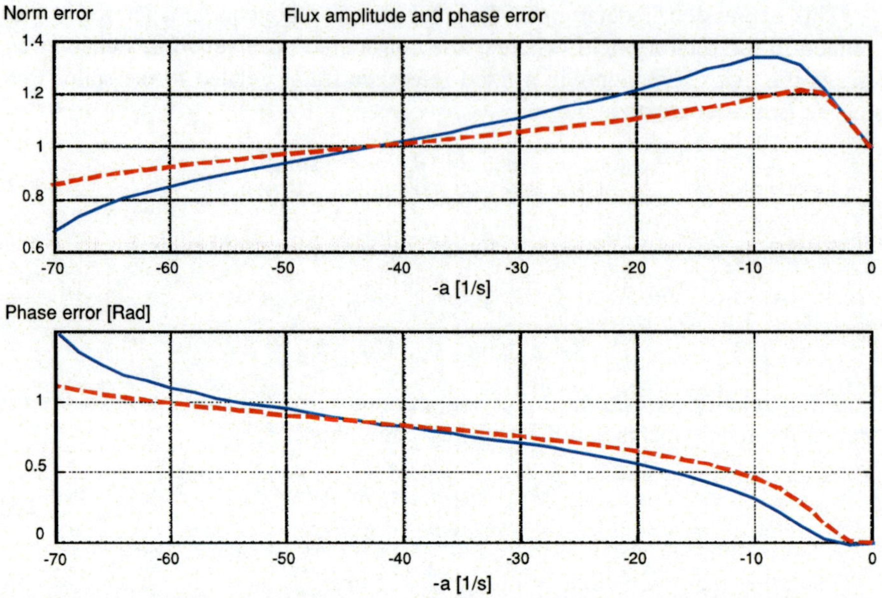

Fig. 4.12 Error on the estimation of the amplitude and phase flux in relation with the real part of the poles when $R_r = 2R_r^*$ (b/a = 0.2; mechanical speed = 0 rpm and electromagnetic torque = 2.3 Nm). *Continuous lines*: without saturation. *Dashed lines*: with saturation

The curves in Fig. 4.12 have been obtained in consideration of a null mechanical speed and a torque close to the rated torque (i.e. 2.3 Nm for the tested machine), as well as a high damping for the observer (i.e. a ratio of the imaginary part onto the real part (*b*/*a*) of the poles whose value is 0.2). The continuous curves have been obtained without taking into account the magnetic saturation, as opposed to the dashed line curves.

Figure 4.12 shows that the estimation error of the flux phase, and consequently the orientation error, increases when the real part of the poles moves away from the origin, and tends toward $\pi/2$. It destabilizes the observer. When the observer gains are null, the real part of the poles is equal to $(-R_r/L_r)$ i.e. for the tested machine: $-11.13\ \mathrm{s}^{-1}$. For practical purposes, the observer poles are chosen in order to increase the observer dynamic. It is the same principle as choosing poles with a real part inferior to $(-R_r/L_r)$, but in this case Fig. 4.12 shows that the observer sensitivity to uncertainties on the rotor resistance is also increased.

Due to the sensitivity study, it is possible to choose gains that strongly reduce the parameter sensitivity of the observer, while getting a satisfactory dynamic. It is also possible to simplify the evolution of gains in relation with the operating point (torque, speed) of the machine. The gains can even become constant on the whole working range, which naturally simplifies the real-time calculation. The choice in gains can also be automated via a software in which different criteria can be introduced:

Choice of gains K

Pole placement method:

Parameter sensitivity:

1. Strongly variable according to the operating point
2. Difficulty to find poles ensuring an acceptable sensitivity for all the operating points

Gains : non-linear functions of speed

Sensitivity study:

Reduced parameter sensitivity

Possibility of fixed gains: simplification

Variable poles according to the operating point

Criteria allowing keeping an acceptable dynamic in a torque and speed range

Fig. 4.13 Comparison of the calculating gain methods

Fig. 4.14 Poles plane

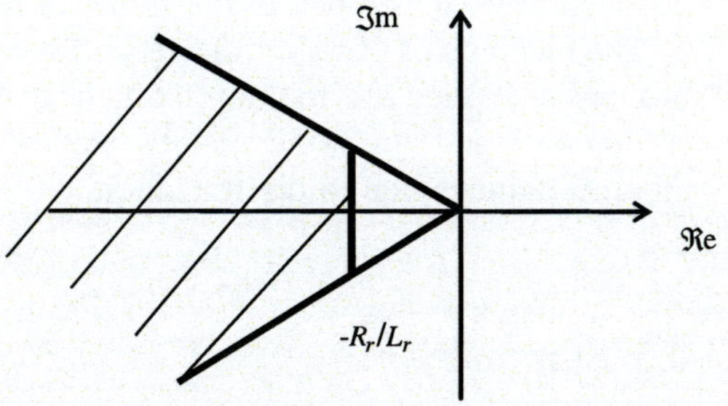

- The real part of the poles ($-a$), in order to ensure rapidity and (most important) stability for the observer;
- The ratio of imaginary part on real part of the poles (b/a), so as to fix the minimum damping required for the flux estimation;
- The error on the flux amplitude;
- The error on the flux orientation of which the value will have to be limited so that the instability risks due to parameters' uncertainties be avoided;
- The variations of the consumed stator current due to parameters uncertainties.

Figure 4.13 compares the classic method for choosing gains by pole placement to the method based on the sensitivity study.

As the method to choose gains is based on the sensitivity study, it is assessed that the observer poles vary, in relation with the operating point of the motor, in a limited range of the poles plane, fixed according to the classic criteria of stability margin (the poles then vary, for instance, in the hatched area introduced at Fig. 4.14). The proposed determining gains method allows obtaining an observer with an optimum adjustment, meaning that the observer is optimized in relation with some criteria. Those criteria can have various natures:

- Dynamic criteria, constraints are imposed on the real and imaginary parts of the poles;
- Sensitivity criteria that limit the acceptable errors on the flux amplitude and orientation;
- Energetic criteria on, for instance, the variations of the consumed stator current;
- Simplicity criteria, the gains must stay constant or can vary in a linear or nonlinear way according to the operating point.

This criteria list is by no means complete. Criteria aiming at minimizing the noise affecting the measured values and criteria taken into account in the Kalman filter can be added to the list. An optimum stochastic observer would be then obtained. The classic deterministic observer (Luenberger observer) takes into account only the dynamic criteria (poles placement).

The considered observers in the rest of the book are all deterministic. They will then all be optimum deterministic observers.

In (Robyns et al. 2000a; Robyns 2000c), the proposed analytic method of parameters sensitivity is applied to a complete-order observer and to a reduced-order observer developed in the frame of reference of the rotating field.

The methods of observer discretization introduced at Sect. 3.5 can also include flux errors if they are too approximate or if the sampling period is too long. In (Delmotte et al. 2001), the analytic method of sensitivity is extended to integrate the uncertainties due to discretization.

4.3 Summary

4.3.1 Control with a Priori on the Rotor Flux

The vector control strategies most frequently used are those orienting the rotor flux. This choice presents the advantage of providing the simplest mathematics expressions. Nonetheless, the stator flux can also be oriented. This solution can be interesting in some cases, as the values relating to the machine stator are more easily accessible than the values relating to the rotor.

From the parameter sensitivity perspective, there is no significant difference between the rotor or stator flux orientation when the axis q component of the flux is assessed null. Indeed, the difference between those two fluxes lies in the leakage fluxes, which are generally very weak. The choice to orientate one or the other will then be linked to the complexity of the mathematics expressions to consider.

Table 4.2 shows the various expressions that allow the determination of the stator pulsation when the rotor flux is oriented (Sect. 3.3.2.4 and 3.3.2.5). The Table 4.3 shows the various estimations of the axis d component of the rotor flux.

Table 4.2 Synthesis of the various techniques for counting the stator pulsation

Explicit autopilot	Implicit autopilot
$\tilde{\omega}_{s1} = p\hat{\Omega} + \frac{R_r}{L_r}\frac{\hat{i}_{sq}}{\hat{i}_\phi}$ (4.38)	$\tilde{\omega}_{s2} = \frac{v_{sq\_reg} - (R_s + \sigma L_s s)\hat{i}_{sq}}{L_s \hat{i}_\phi}$ (4.39)

Table 4.3 Synthesis of the various flux estimators of the rotor flux (ϕ_{rd})

Estimation using the current	Estimation using the voltage
$\tilde{i}_{\phi 1} = \frac{1}{1+\tau_r s}\hat{i}_{sd}$ (4.43)	$\tilde{i}_{\phi 2} = \frac{1}{R_s}\frac{v_{sd\_reg} + \sigma L_s \omega_s \hat{i}_{sq}}{\sigma \tau_s \tau_r s^2 + (\tau_s + \tau_r)s + 1}$ (4.44)

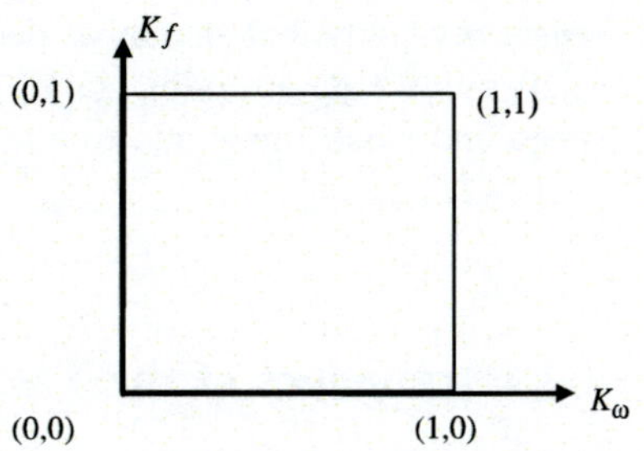

Fig. 4.15 Plan (K_f, $K\omega$) limited by four different control algorithms

As both methods to calculate $\tilde{i}_{\phi r}$ and calculate $\tilde{\omega}_s$ present advantages as well as disadvantages from the parameter sensitivity perspective, it is proposed to calculate $\tilde{i}_{\phi r}$ and $\tilde{\omega}_s$ using a combination of, on the one hand (4.43) and (4.44), and on the other hand (4.38) and (4.39), such as:

$$\begin{aligned} \tilde{i}_{\phi r} &= (1 - K_f)\,\tilde{i}_{\phi 1} + K_f\,\tilde{i}_{\phi 2} \quad \text{avec} \quad 0 \le K_f \le 1 \\ \tilde{\omega}_s &= (1 - K_\omega)\,\tilde{\omega}_{s1} + K_\omega\,\tilde{\omega}_{s2} \quad \text{avec} \quad 0 \le K_\omega \le 1 \end{aligned} \tag{4.58}$$

With coefficients K_f and K_ω, there are four control algorithms which are combined, corresponding to extreme values of those coefficients: $(K_f, K_\omega) = (0,0)$, $(K_f, K_\omega) = (0,1)$, $(K_f, K_\omega) = (1,0)$ and $(K_f, K_\omega) = (1,1)$. This formulation is the basement of a multialgorithmic control

The case $(K_f, K_\omega) = (0,0)$ corresponds to the most classical control strategy, whereas the case $(K_f, K_\omega) = (1,1)$ was proposed in (Caron and Hautier 1995) in the context of an oriented stator flux control. The combination of the cases $(K_f, K_\omega) = (0,0) and (K_f, K_\omega) = (0,1)$ was proposed in (Robyns et al. 1998, 2000b).

The assessment of various algorithms and the evolution of one algorithm toward another, according to the operating point torque-speed, can be realized thanks to the fuzzy logic. Coefficients (K_f, K_ω) can a priori take all the values included in the square represented in Fig. 4.15. A theoretical sensitivity study, taking into account the magnetic saturation, allows optimizing the choice of coefficients (K_f, K_ω) for all the operating points so as to minimize the parameter sensitivity. The fuzzy logic allows obtaining progressive and nonlinear evolutions of the coefficients (K_f, K_ω), and the easy integration of the information deduced not only from the theoretical sensitivity study, but also from simulations and experimentations.

4.3.2 Control Without a Priori

Associating the fuzzy logic with the theoretical analysis of sensitivity in order to optimize the choice of observer gains for all the operating points is also interesting. The fuzzy logic allows manual integration of the information deduced from the sensitivity study, but also those deduced from simulations and experimentations. It easily allows determining gains evolving in a nonlinear way with mechanical speed and electromagnetic torque while integrating various criteria developed in Sect. 4.2.3.2.

4.4 Parameters of the Tested Machine

Rated power: 750 W
Rated voltage: 220 V (star coupling)
Number of poles: 2
Stator resistance: 3.5 Ω
Rotor resistance: 1.8 Ω
Mutual inductance: 154 mH
Stator inductance: 160 mH
Rotor inductance: 160 mH
β: 0.78
s: 8.8

References

Robyns, B., Labrique, F., & Buyse, H. (1996). Commande numérique simplifiée et robuste d'actionneurs asynchrones de faible puissance. *Journal de Physique III, 6*, 1039–1057. (August 1996).

Hautier, J.P., & Caron, J.P. (1999). Convertisseurs statiques: méthodologie causale de modélisation et de commande. Paris: Editions Technip, ISBN 2–7108-0745–9

Robyns, B., Fu, Y., Buyse, H., & Labrique, F. (1994). *Flux control performance of an induction motor indirect F.O.C using a simplified current control strategy*: *Proceedings of ICEM'94*, Paris, Sept 1994 (Vol. 2, pp. 374–379)

Robyns, B., Sente, P., Labrique, F., & Buyse, H. (1999). Influence of digital current control strategy on the sensitivity to electrical parameter uncertainties of induction motor indirect field oriented control. *IEEE Transactions on Power Electronics, 49*(4), 690–699. (juillet 1999).

Robyns, B., Meuret, R., & Sente, P. (2002). Commande vectorielle indirecte de la machine asynchrone. *Influence de la stratégie de régulation digitale des courants sur les performances de la commande, Revue Internationale de Génie Electrique, Hermès, 5*(1), 35–47.

Robyns, B., Berthereau, F., Cossart, G., Chevalier, L., Labrique, F., & Buyse, H. (2000a). A methodology to determine gains of induction motor flux observers based on a theoretical parameter sensitivity analysis. *IEEE Transactions on Power Electronics, 15*(6), 983–995. (Nov 2000).

Robyns, B. (2000c). Synthèse de commande robustes pour machine asynchrone basée sur une théorie caractérisant la sensibilité paramétrique. Research report from Université des Sciences et Technologies de Lille

Delmotte, E., Semail, B., Robyns, B., & Hautier, J. P. (2001). Flux observer for induction machine control. Part I: Sensitivity analysis as a function of sampling rate and parameters variations. Part II: Robust synthesis and experimental implementation. *The European Physical Journal, Applied Physics, 14,* 13–43.

Caron, J.P., & Hautier, J.P. (1995). Modélisation et commande de la machine asynchrone. Paris: Editions Technip, ISBN 2–7108-0683–5

Robyns, B., Buyse, H., & Labrique, F. (1998). Fuzzy logic based field orientation in an indirect FOC strategy of an induction actuator. *Mathematics and Computer in Simulation, 46,* 265–274.

Robyns, B., Berthereau, F., Hautier, J. P., & Buyse, H. (2000b). A fuzzy logic based multimodel field orientation in an indirect F.O.C. of an induction motor. *IEEE Transactions on Industrial Electronics, 47*(2), 380–388. (avril 2000).

Chapter 5
Fuzzy Supervisor

The fuzzy logic aims at translating linguistic rules, that are called fuzzy rules, into a mathematical form. It also aims at describing the observations and reactions a human operator would have during a process control. The results deduced from fuzzy logic are fully deterministic. The fuzzy notion needs to be linked to the uncertainty principle that can be found in most of the systems we use in practice. This uncertainty is mathematically formalized by a membership function.

5.1 Fuzzy Logic Principles

In the former chapter, we noticed that some controls were highly sensitive to rotor resistance uncertainties, especially when the mechanical speed is too big. We also noticed that other controls where highly sensitive especially when the mechanical speed and the electromagnetic torque are small. This brings up the fuzzy notions of big and small, that are currently being used in practice. The fuzzy logic not only enables to mathematically formalize these fuzzy notions but also to deduce precise actions from the fuzzy rules based on our findings.

Three main steps can be distinguished in a fuzzy reasoning: fuzzification, inference and defuzzification (Buhler 1994; Borne et al. 1998).

Fuzzification consists in associating a particular fuzzy subset with a given variable's value (speed, torque, etc.). Linguistic variables are therefore used, variables that are mathematically represented by membership functions, that quantify the relative uncertainties of the variable belonging to this set. Those linguistic variables are chosen so as to modelize the observations of a human being, who will qualify a phenomenon as positive, negative, null, big, small or medium. This set constitutes a universe of discourse. Figure 5.1 shows the three membership functions that enable to qualify the speed variable amplitude for the three fuzzy subsets. S stands for the set of small values, M for the set of medium ones and B for the set of big ones. The first step of fuzzification consists in

B. Robyns et al., *Vector Control of Induction Machines*, Power Systems,
DOI: 10.1007/978-0-85729-901-7_5, © Springer-Verlag London 2012

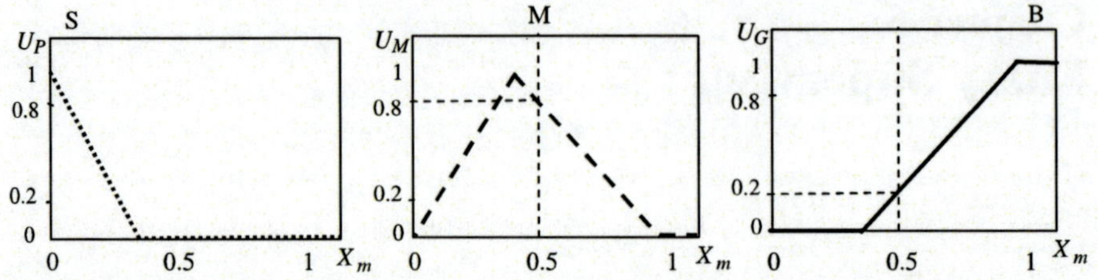

Fig. 5.1 Examples of speed variable's membership functions S = Small, M = Medium and B = Big

normalizing the input quantities while using scaling factors. In a second step, the expert has to define a universe of discourse to qualify each quantity. This universe of discourse has to be made up of a finite number of fuzzy subsets. To each of them corresponds a membership function. If input value X_m (speed normalized value) equals 0.5, Fig. 5.1 shows us that the membership factor of this X_m value to fuzzy set M (medium) equals $U_M(X_M) = 0.8$ and that its membership factor to fuzzy set B (big) equals $U_B(X_M) = 0.2$. As a general rule, the generated value reflects the variable's membership degree to its corresponding fuzzy subset.

Inference links fuzzified input quantities to the output variable, that is also expressed in linguistic form, following a certain number of fuzzy rules. The fuzzy rules are expressed using the IF conditional term and the THEN deductive term. They enable to link conditions to the conclusion. If a rule has several conditions, they are linked either by the AND operator or by the OR operator. Here is an example of fuzzy rule, where X_r and X_m correspond to the fuzzified input quantities and X_k corresponds to the fuzzified output quantity.

- if X_r is big (B) and X_m is small (S) then X_k is big (B);
- if X_r is small (S) or X_m is big (B) then X_k is small (S).

Each rule enables to deduce the fuzzy set to which the output quantity belongs. The membership function of this set then has to be determined. The inference method enables to do so by determining the realization of the various logical operators. Various inference methods can be used. The following part of this chapter will deal with the sum-product inference method (Buhler 1994). For this method, the OR operator is realized at the level of condition by the form of the sum, more precisely by the medium value, whereas the AND operator is realized by the form of the product. The result is a global membership factor obtained when the conditional term, written as U_c, is applied. The conclusion of each rule, preceded by THEN, is realized when forming the product of the membership factor of the condition with the membership function of the output quantity. The result obtained from this rule is a membership function weighted to the fuzzy set of output quantity (U_R). A numerical value will then have to be extracted from it. From the evaluation of each rule, the conclusion is obtained when applying the OR operator, that is implemented by the sum of the outputs for each rule.

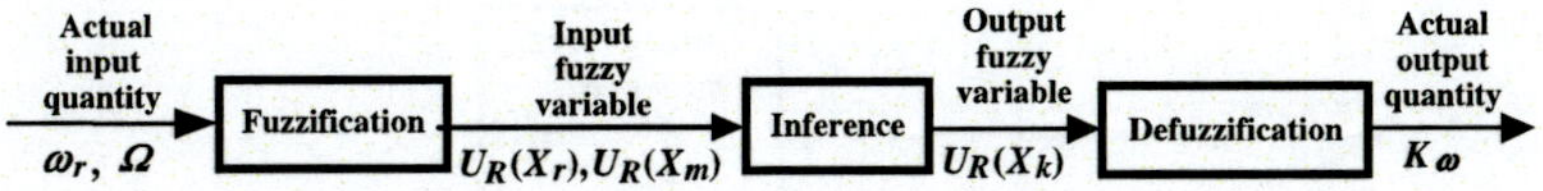

Fig. 5.2 Sequence of operations characteristic of fuzzy logic

An application example of this method can be found in the following part of this chapter (Fig. 5.7).

Defuzzification consists in transforming fuzzy information deriving from inference into deterministic information that can be directly applied to the process. The most commonly used defuzzification method is the one that determines the center of gravity of the resulting belonging function (Buhler 1994), that can be calculated while using the relation (5.1):

$$X_R = \frac{\int_{-1}^{1} X_k U_R(X_k) \mathrm{d}X_k}{\int_{-1}^{1} U_R(X_k) \mathrm{d}X_k} \tag{5.1}$$

The integral in the denominator gives the surface, while the integral in the numerator corresponds to the surface moment. This method will be applied to a concrete case in the following section. Figure 5.2 sums up the sequence of operations characteristic to the fuzzy logic.

5.2 Combination of Two Flux Orientation Strategies

5.2.1 Combination of Explicit and Implicit Techniques for Frame Orientation

Since both methods (4.38) and (4.39) enable to determine the stator pulsation, both ω_s have advantages and drawbacks. The following of this chapter shows how important it is to combine them. The stator pulsation is then determined by the following expression:

$$\tilde{\omega}_s = (1 - K_\omega)\,\tilde{\omega}_{s1} + K_\omega\,\tilde{\omega}_{s2} \quad 0 \leq K_\omega \leq 1 \tag{5.2}$$

The values of parameter K_ω are determined from the theoretical study of sensitivity. They are then refined using simulations and experimental tests, in order to reduce the control sensitivity to the various parametric uncertainties. Parameter K_ω must then vary together with the machine operating point. The fuzzy logic is particularly suitable to determine the value of K_ω, since it consists in dealing with

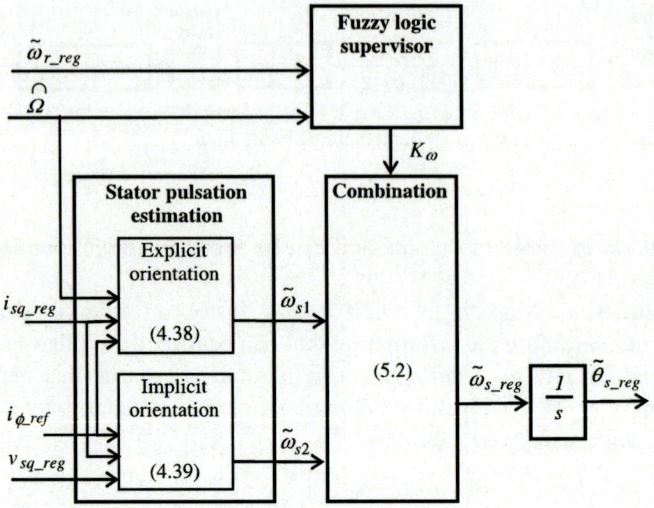

Fig. 5.3 Technique for frame orientation using a fuzzy logic supervisor

uncertainties. Figure 5.3 pictures the global structure of the technique for frame orientation, including a fuzzy logic supervisor used to calculate parameter K_ω. This set can then easily be inserted into the vector control with a priori on the orientation of the rotor-flux (Fig. 3.28).

The strategy of torque control is based on indirect flux control, pictured in Fig. 4.3. The current controllers mentioned are PI-type controllers. To determine the two estimations of the stator pulsation, the current i_{sq_reg} is used (Robyns et al. 1998, 2000b).

5.2.2 Variations in the Consumption of Stator Current

The variations in the consumption of stator current due to poor flux control can be determined from expressions (4.4), (4.34) and (4.37). Figure 5.4 shows the increase in the consumption of stator current by the machine in steady state. This increase results from a 100% error on the rotor resistance ($R_r = 2R_r^*$), depending on K_ω, when the mechanical speed equals 1,500 rpm (rotations per minute) and when the torque equals 2.3 Nm. The curve in continuous line is deduced from experimental measurements. It confirms the theoretical predictions pictured by the curve in dotted lines, obtained when considering the magnetic saturation. This figure pictures how system sensitivity to errors on R_r sharply decreases for small values of K_ω. Therefore, it is not necessary to choose a value of K_ω close to the unit, to avoid the risks of instability mentioned in Sect. 4.2.2.2.

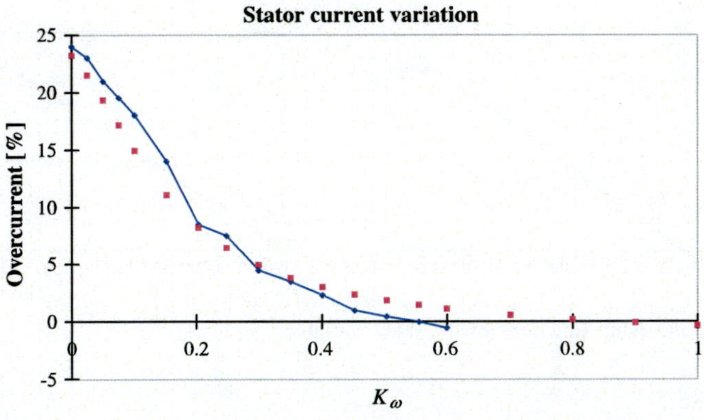

Fig. 5.4 Increase in the consumption of stator current depending on K when $R_r = 2R_r^*$ (speed = 1,500 rpm and torque = 2.3 Nm). Curve in *dotted line*: theoretical values. Curve in *continuous line*: experimental values

5.2.3 Introduction of a Fuzzy Logic Supervisor

5.2.3.1 Development of a Fuzzy Logic Supervisor

Values for K_ω are naturally chosen so as to reduce the flux control sensitivity to the uncertainties on the rotor resistance. However, the theoretical study enabled to highlight the fact that K_ω had to be chosen in order to avoid a too important sensitivity to inductances. Furthermore, one must take into account the fact that implicit technique for frame orientation (4.39) relies on the stator resistance uncertainties. K_ω will then depend on the mechanical speed and on slip pulsation ω_r, deduced from (3.67), that reflects the electromagnetic torque.

Three steps can be highlighted when determining K_ω using fuzzy logic: fuzzification, inference and defuzzification. For the case studied, these three steps can be the following:

Fuzzification

Fuzzification consists in defining the membership functions for the various variables, particularly for the input variables. The transformation of physical quantities (determined quantities) into linguistic quantities (fuzzy variables), that can be treated by inference, can then be realized. In this particular case, the linguistic variables, also called normalized variables, are defined by:

$$X_r = \left|\frac{\omega_r}{\hat{\omega}_r}\right| \quad X_m = \left|\frac{\Omega}{\hat{\Omega}}\right| \tag{5.3}$$

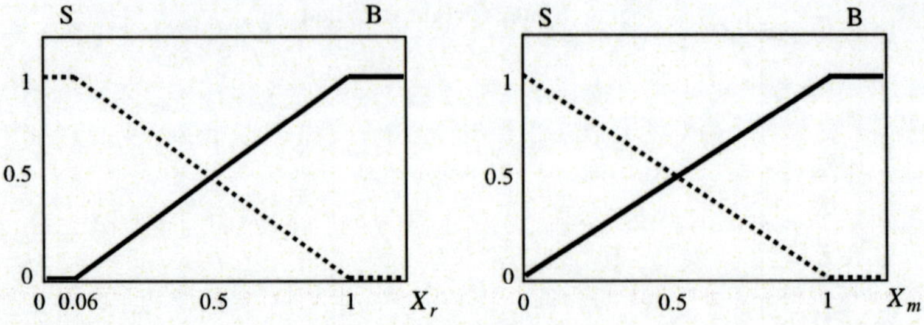

Fig. 5.5 Membership functions of the input variables

Table 5.1 States of output variables depending on the sensitivity to R_r

		X_r	
	X_k	S	B
X_m	S	X	B
	B	X	B

Table 5.2 States of output variables depending on the sensitivity to R_s

		X_r	
	X_k	S	B
X_m	S	S	S
	B	X	X

Table 5.3 States of output variables depending on the sensitivity to M

		X_r	
	X_k	S	B
X_m	S	S	S
	B	B	B

Symbol ^ stands for the maximum value of the corresponding quantity. The choice of membership functions for normalized variables is pictured in Fig. 5.5. For those two-state variables, we can limit ourselves to the following: big (B) and small (S).

Table 5.4 States of output variables depending on the dynamic factors

		X_r	
	X_k	S	B
X_m	S	X	X
	B	S	X

Table 5.5 Synthesis of Tables 5.1, 5.2, 5.3 and 5.4

		X_r	
	X_k	S	B
X_m	S	S	B
	B	S	S

Inferences

The inference is determined by the laws linking input variables to output variables. To define these laws, a sensitivity study is carried out for each parameter and determination method of ω_s. Tables 5.1, 5.2 and 5.3 can be deduced from these studies. X_k represent the output value and is the value that will be taken by K_ω. These tables present the state the output variable should be in for the sensitivity of R_r, R_s and M to be reduced. X stands for any state. Simulations are carried out for each case taking dynamic factors into account. Table 5.4 is then drawn up. Tables 5.1, 5.2, 5.3 and 5.4 show contradictory tendencies.

Preferential choices are necessary for the synthesis of Table 5.5. Such choices depend on the precise knowledge we have for each parameter or on the degree of preference for each operating point. The priority for synthesis Table 5.5 was to reduce the sensitivity to R_r for the most sensitive operating points, those with a low speed and a high torque. The conclusions of Fig. 5.4 are also taken into account, as they enable us to keep the value of K_ω low and thus avoid the instability issues raised in Sect. 4.2.2.2. Laws can be deduced from Table 5.5 and tested experimentally. Depending on the real-time results to these experiments, the laws are either considered true or amended. The following laws were deduced from Table 5.5:

- if X_r is big (B) and X_m is small (S) then X_k is big (B);
- if X_r is small (S) or X_m is big (B) then X_k is small (S).

When $\omega_r\Omega<0$, K_ω is equal to zero. The first law aims at reducing the sensitivity of the flux control to uncertainties on R_r, because the sensitivity is usually at

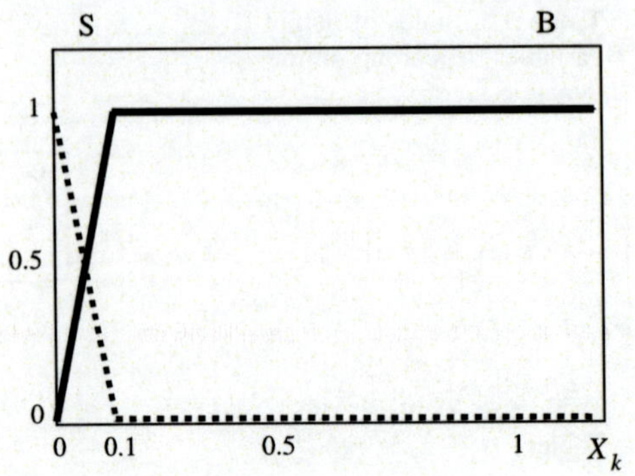

Fig. 5.6 Membership function of the output variable

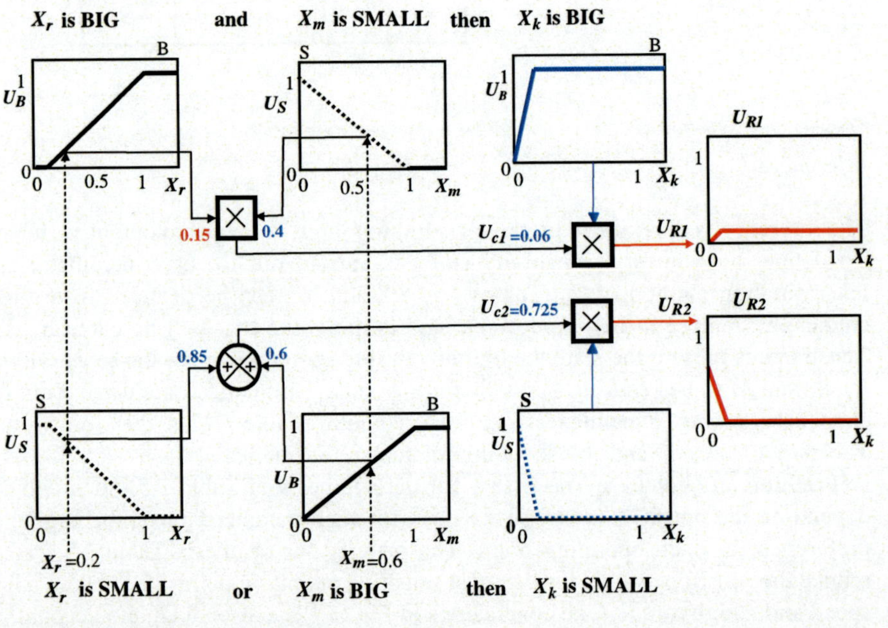

Fig. 5.7 Application of the inference

its highest for a high torque and a low speed. The second law aims at reducing this sensitivity to inductances, because, as shown in (4.42), it is high for a low torque and a high speed. The second law also makes it possible to reduce the sensitivity to uncertainties on R_s, as it is at its highest at zero speed. The methodology used here is the sum-product inference (Buhler 1994), although other methods could be used (max–min inference, max–product inference etc.).

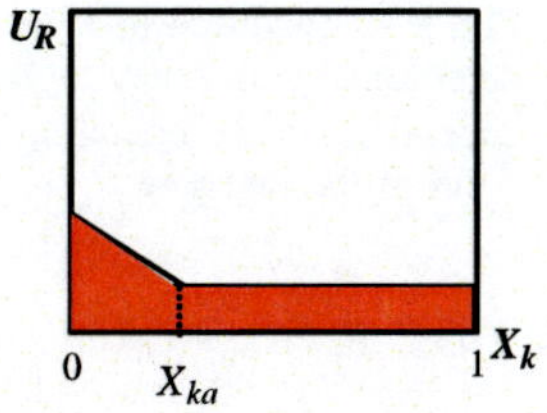

Fig. 5.8 Extraction of the membership function to the output set

Defuzzification

Inferences provide fuzzy information. Defuzzification allows for the translation of the fuzzy variable into a specific control signal. Figure 5.6 shows the membership function of the output variable X_k. The most common defuzzification method consists in determining the center of gravity of the resulting membership function. That makes it possible to determine the value of K_ω.

A concrete example of the processing of the proposed fuzzy algorithm is developed in Fig. 5.7. It is assumed that the normalized input values are: $X_r = 0.2$ and $X_m = 0.6$.

The first rule, which reads "If X_r is big and X_m is small, then X_k is big", calls for the membership functions to be evaluated:

- membership of the variable X_r to the "big" set (B): $U_B(X_r = 0.2) = 0.15$,
- membership of the variable X_m to the "small" set (S): $U_S(X_m = 0.6) = 0.4$.

Applying the conditional term "and" gives the value $U_{C1} = 0.15 \times 0.4 = 0.06$.

The deduction term "then" is evaluated through balancing of the output variable membership function to the set "big" with this value. Applying this rule leads to an intermediate membership function: $U_{R1}(X_k) = 0.06 \times U_B(X_k)$.

The second rule, "If X_r is small and X_m is big then X_k is small" calls for the membership functions to be evaluated:

- membership of the variable X_r to the "small" set (S): $U_S(X_r = 0.2) = 0.85$,
- membership of the variable X_m to the "big" set (B): $U_B(X_m = 0.6) = 0.6$.

The conditional term of the second rule, "or", is evaluated through calculation of the average value: $U_{C2} = (0.85 + 0.6)/2 = 0.725$.

The deduction term "then" of the second rule is evaluated through balancing of the membership function of its output variable to the set "small" with this value: $U_{R2}(X_k) = 0.725 \times U_S(X_k)$.

The resulting membership function, shown in black on Fig. 5.8, is obtained through extraction of the average value of the two partial membership functions:

$$U_R(X_k) = \frac{1}{2}(U_{R1}(X_k) + U_{R2}(X_k)) = \frac{1}{2}(U_{C1}U_B(X_k) + U_{C2}U_S(X_k)) \quad (5.4)$$

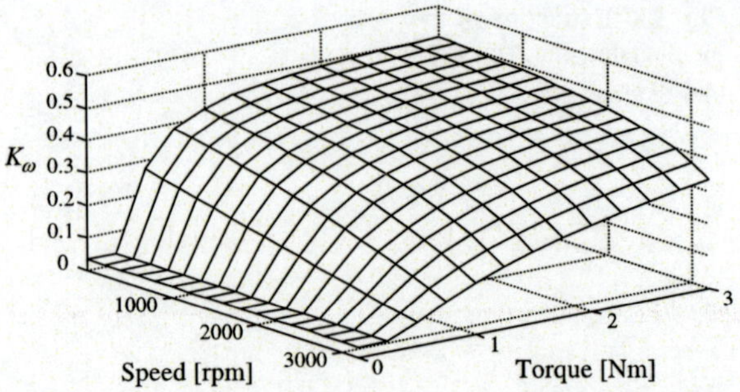

Fig. 5.9 Evolution of K as determined with fuzzy logic, depending on the operating point of the machine

$U_R(X_k)$ is a fuzzy information, from which determinist information applicable to the control must be deduced. The most common defuzzification method consists in determining the center of gravity of the resulting membership function $U_R(X_k)$. Applying relation (5.1) to the case studied in this section, gives us:

- for the moment of the surface (numerator):

$$M = \frac{1}{2}\left[U_{C1}\int_0^1 X_k U_B(X_k)\,\mathrm{d}X_k + U_{C2}\int_0^1 X_k U_S(X_k)\,\mathrm{d}X_k\right] \quad (5.5)$$

$$\text{that is to say } M = \frac{1}{2}\left[U_{C1}\left(\frac{1}{2} - \frac{X_{\mathrm{ka}}^2}{6}\right) + U_{C2}\left(\frac{X_{\mathrm{ka}}^2}{6}\right)\right]$$

- for the surface (denominator):

$$S = \frac{1}{2}\left[U_{C1}\int_0^1 U_B(X_k)\,\mathrm{d}X_k + U_{C2}\int_0^1 U_S(X_k)\,\mathrm{d}X_k\right] \quad (5.6)$$

$$\text{that is to say } S = \frac{1}{2}\left[U_{C1}\left(1 - \frac{X_{\mathrm{ka}}}{2}\right) + U_{C2}\left(\frac{X_{\mathrm{ka}}}{2}\right)\right]$$

The value of the output variable is consequently (($X_{\mathrm{ka}} = 0.1$, $U_{C1} = 0.06$ and $U_{C2} = 0.725$) :

$$X_R = \frac{M}{S} = \frac{0.0155}{0.0466} = 0.334 \quad (5.7)$$

Lastly, we obtain the value of coefficient $K_\omega = X_R$.

Figure 5.9 shows the evolution of K_ω, as deduced with fuzzy logic, depending on the operating point of the machine. Fuzzy logic makes it possible to determine a non-linear two-variable function for K_ω quite easily.

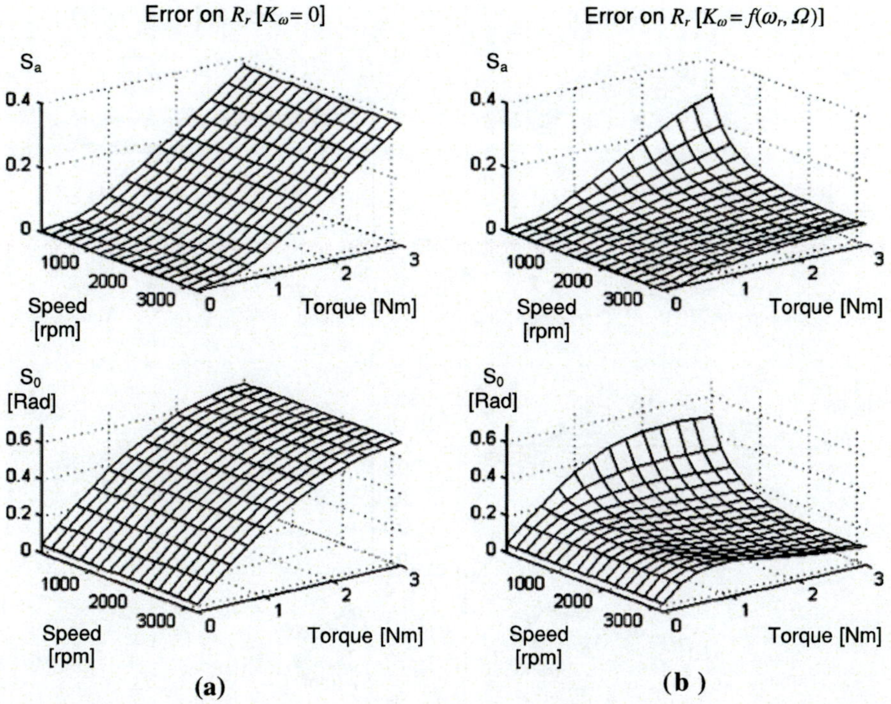

Fig. 5.10 Sensitivity of amplitude and flux orientation to uncertainties on R_r. **a** $K = 0$, **b** K is determined through fuzzy logic

5.2.3.2 Theoretical Comparison

To achieve the theoretical sensitivity analysis, coefficients A_2, B_2 and D in equation (4.5) can be determined as follows:

$$\begin{aligned} A_2 &= (1 - K_\omega)\,A_2' + K_\omega\,A_2'' \\ B_2 &= (1 - K_\omega)\,B_2' + K_\omega\,B_2'' \\ D &= (1 - K_\omega)\,D' + K_\omega\,D'' \end{aligned} \tag{5.8}$$

Coefficients A_2', B_2' and D' are equivalent to A_2, B_2 and D deduced from Eq. 4.27, whereas A_2'', B_2'' and D'' are deduced from Eq. 4.40. Coefficients A_1, B_1 and C are determined with the help of Eq. 4.27.

Figures 5.10, 5.11 and 5.12 show the sensitivity of amplitude and flux orientation to errors on, respectively, R_r, R_s and M, depending on the mechanical speed and the electromagnetic torque. A 1% error is introduced on the relevant parameter. These figures make a comparison of the control sensitivity possible for $K_\omega = 0$ (curves A) and for a value of K_ω determined through fuzzy logic (curves B).

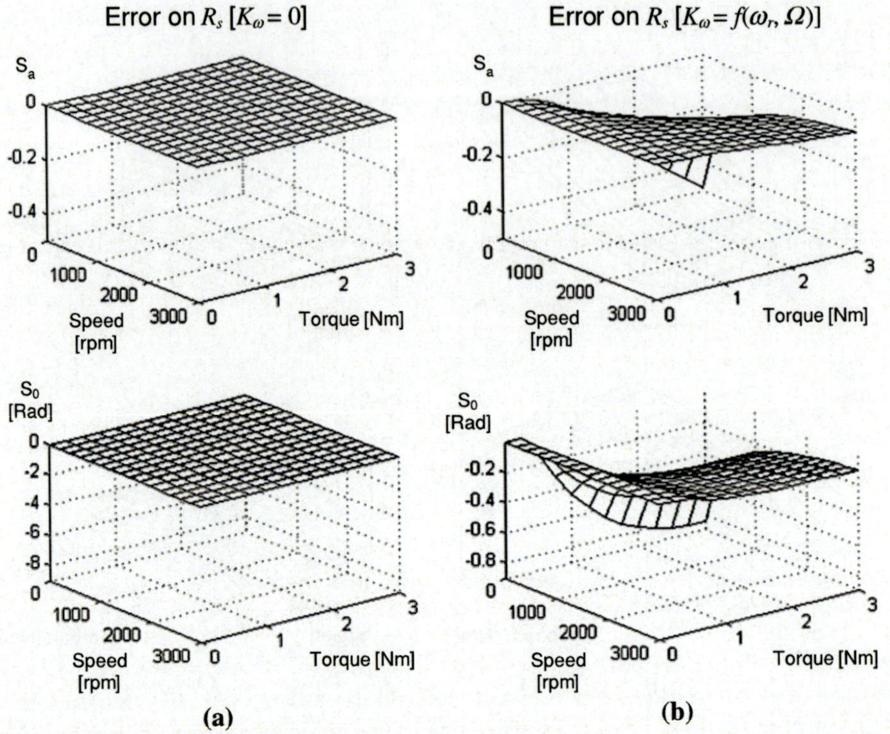

Fig. 5.11 Sensitivity of amplitude and flux orientation to uncertainties on R_s **a** $K = 0$, **b** K is determined through fuzzy logic

Figure 5.10 shows clearly that a multi-model control allows for a dramatic drop in control sensitivity to the rotor resistance. However, this improvement is somewhat impaired by the introduction of a sensitivity to the stator resistance, although this sensitivity only appears at some operating points (see Fig. 5.11).

Figure 5.12 shows that in the event of uncertainties on inductances, the multi-model control results in a decreased sensitivity of the flux amplitude but in an increased sensitivity of the flux orientation [see (4.42)]. This sensitivity remains nevertheless acceptable: although Fig. 5.12b shows that a 100% error on M can result in an orientation error of over 2 rad for some operating points, which would destabilize the system. In practice, the inductance M is usually known to a precision of a few percents. If for example the error on M is divided by 10 and thus equal to 10%, the maximum orientation error is divided by the same factor and drops to 0.2 rad, a much more acceptable value.

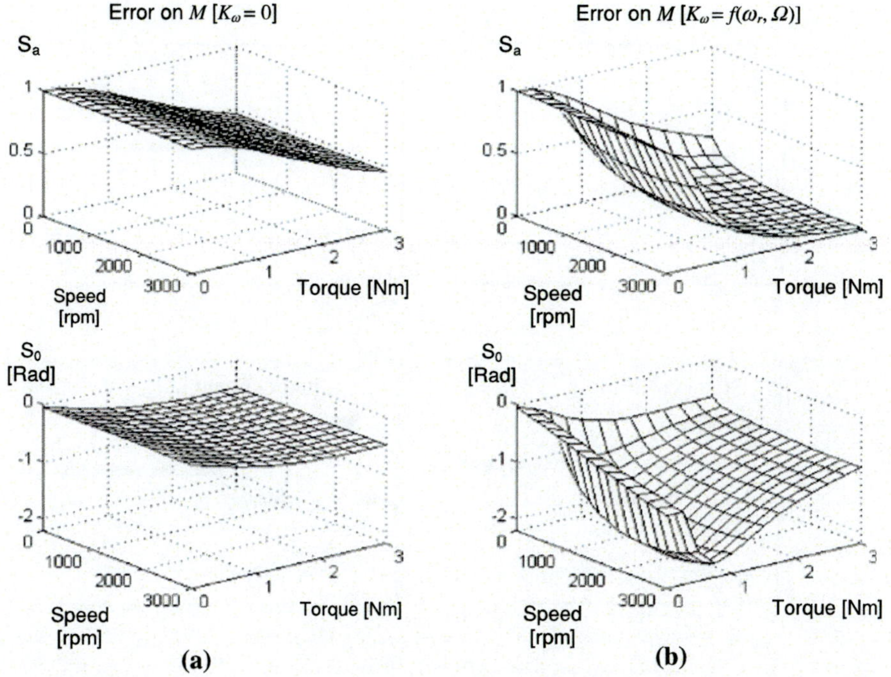

Fig. 5.12 Sensitivity of amplitude and flux orientation to uncertainties on M (not taking magnetic saturation into account); **a** $K = 0$, **b** K is determined through fuzzy logic

5.2.3.3 Experimental Comparison

Figures 5.13 and 5.14 show the experimental response and the reference of the speed, the response of the RMS (root mean square) stator current and the response of K to a speed range of 0–1,500 rpm, followed by a load torque range of 2.3 Nm. Tests shown in Fig. 5.13 are carried out with optimized parameters, whereas an error of 100% on R_r is introduced in tests shown in Fig. 5.14. Comparison of Fig. 5.14a and b confirms that combining several models allows for a much lower sensitivity of the flux control to uncertainties on R_r.

5.3 Combination of Two Strategies for Flux Estimation

5.3.1 Combination of the Magnetizing Current Estimations

In the previous section, we have shown how combining two stator frequency estimation methods improves precision. We will now use the same approach to demonstrate the use of a combination of two magnetizing current estimators (Table 4.3) (Robyns et al. 1997; Berthereau et al. 2001a; Berthereau 2001b).

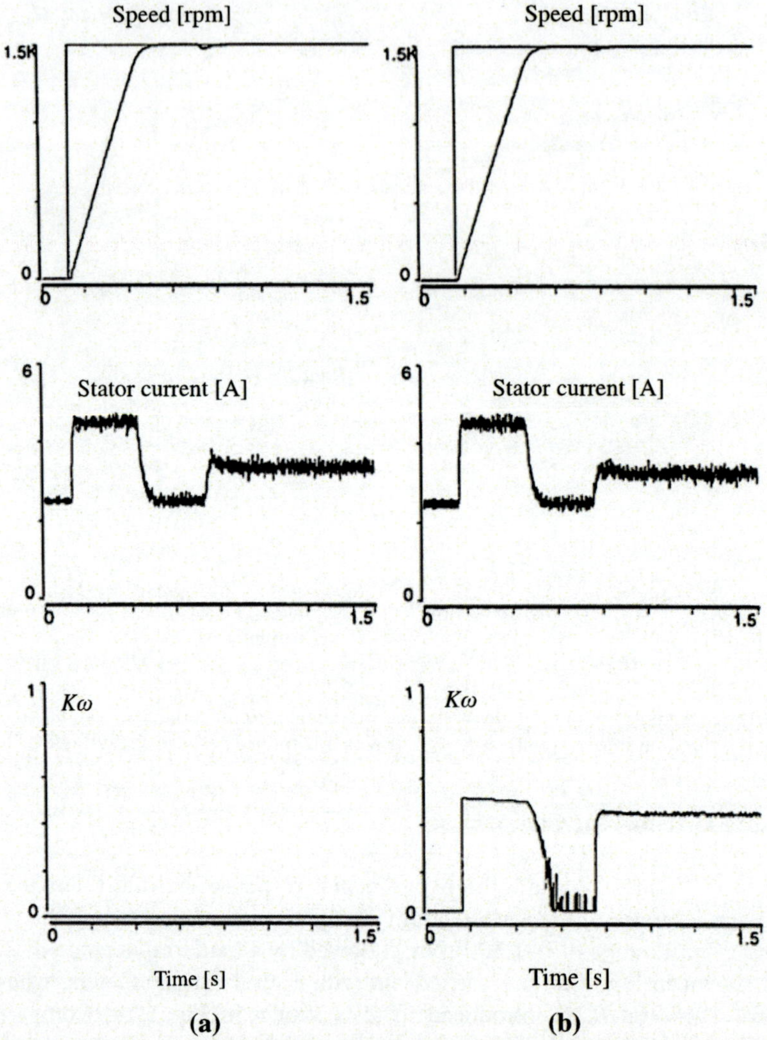

Fig. 5.13 Experimental response to a speed step followed by a load torque step, **a** $K = 0$, **b** K is determined through fuzzy logic

Combination (5.8) then determines the estimation of magnetizing current and must be added to expression (5.2) (below):

$$\tilde{i}_{\phi r} = (1 - K_f)\tilde{i}_{\phi 1} + K_f \tilde{i}_{\phi 2} \quad 0 \leq K_f \leq 1 \tag{5.9}$$

$$\omega_s = (1 - K_\omega)\,\omega_{s1} + K_\omega\,\omega_{s2} \quad 0 \leq K_\omega \leq 1 \tag{5.2}$$

As it was the case for coefficient K_ω, the values of coefficient K_f are determined from the theoretical study of sensitivity. They are then refined using simulations

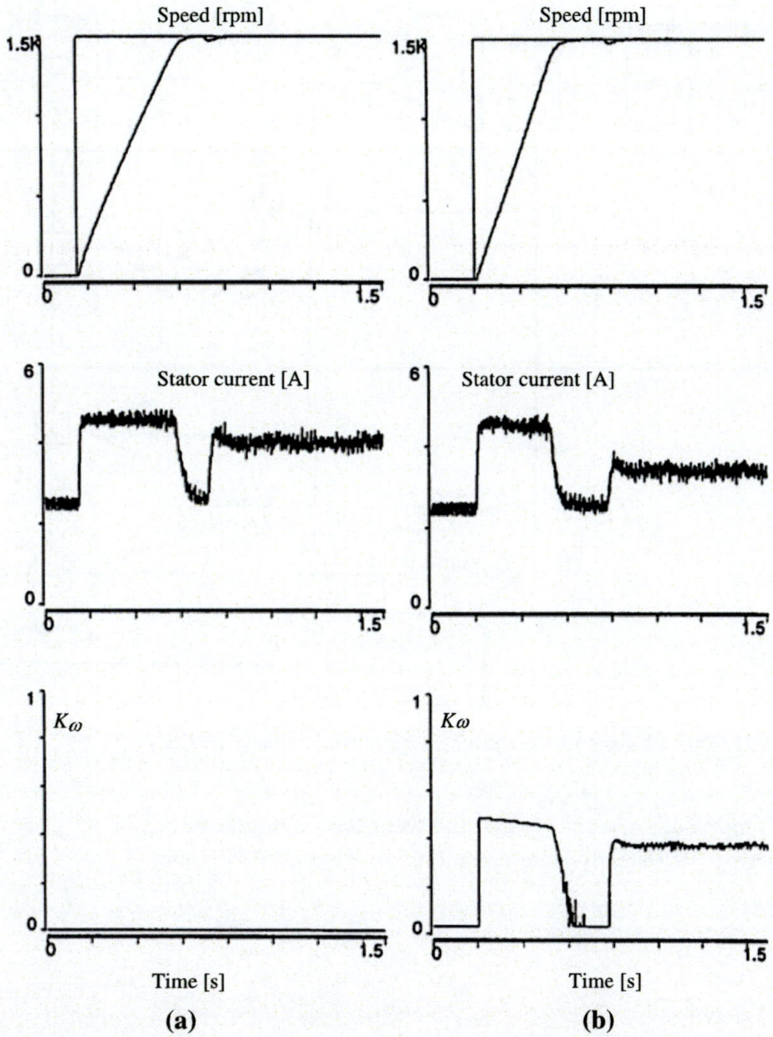

Fig. 5.14 Experimental response to a speed step followed by a load torque step with $R_r = 2R_r^*$, **a** $K = 0$, **b** K is determined through fuzzy logic

and experimental tests, in order to reduce the control sensitivity to parametric uncertainties.

Fuzzy logic is also used to determine K_f depending on the operating point.

The control strategy studied in this section is a direct vector control with explicit orientation of the rotor-flux (Fig. 3.27). A proportional-integral controller is used. Rather than on measures of the currents, the decoupling is based on their references for the calculation of the estimated EMFs (electromotive forces) (Fig. 5.15).

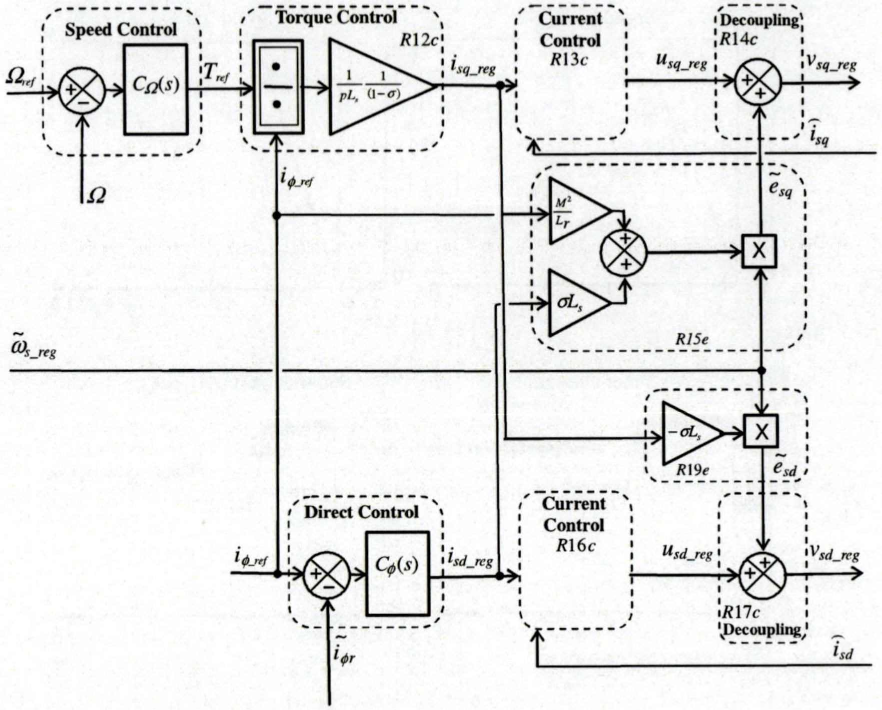

Fig. 5.15 Direct control using adjustment quantities for compensations

Compensation supposes that the rotor-flux is oriented $\left(\phi_{\mathrm{rd}} = \phi_{r\_\mathrm{ref}} = Mi_{\phi r} = Mi_{\mathrm{sd}}, \phi_{\mathrm{rq}} = 0\right)$; steady state equations (4.25a and b) become:

$$v_{\mathrm{sd}} = R_s\, i_{\mathrm{sd}} - \sigma\, L_s\, \omega_s\, i_{\mathrm{sq}} = u_{\mathrm{sd}} - \sigma\, L_s\, \omega_s\, i_{\mathrm{sq}} \tag{5.10}$$

$$v_{\mathrm{sq}} = R_s\, i_{\mathrm{sq}} + \sigma\, L_s\, \omega_s\, i_{\mathrm{sd}} + \frac{M^2}{L_r} i_{\phi r}\omega_s = u_{\mathrm{sq}} + \sigma\, L_s\, \omega_s\, i_{\mathrm{sd}} + \frac{M^2}{L_r} i_{\phi r}\omega_s \tag{5.11}$$

By analogy with Eqs. 3.61 and 3.64, the estimate of the quadrature EMF can be simplified to:

$$R15e \rightarrow \tilde{e}_{\mathrm{sq}} = \left(\sigma\, L_s\, i_{\mathrm{sd\_ref}} + \frac{M^2}{L_r} i_{\phi r\_\mathrm{ref}}\right)\omega_{\mathrm{s\_reg}} \tag{5.12}$$

Contrary to the diagram shown in Fig. 3.28, frame orientation is realized with a fuzzy supervisor (Fig. 5.3); an estimator of the magnetizing current with a fuzzy supervisor is also used (Fig. 5.16).

Fig. 5.16 Estimation of the magnetizing current using a fuzzy logic supervisor

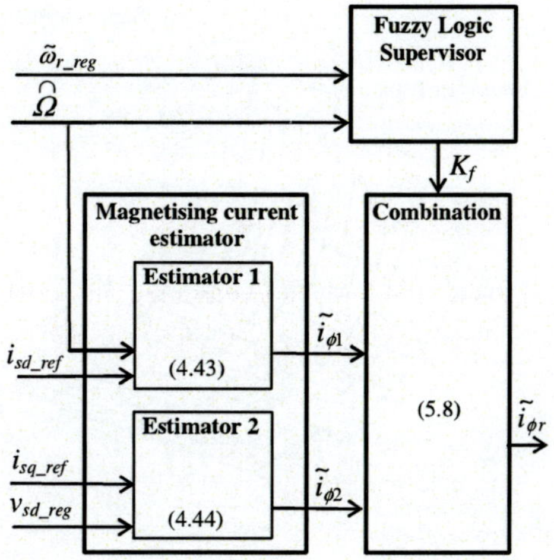

5.3.2 Theoretical Constraint

For $(K_f, K_\omega) = (0, 0)$, the control sensitivity is as usual. Flux sensitivity to uncertainties on the rotor resistance is identical to that of Figs. 4.6a and 5.10a. This sensitivity is very high.

For $(K_f, K_\omega) = (0, 1)$, the error on flux orientation corresponds to expression (4.41). This means that, for $\omega_r = 0$, when there is an error on inductance L_s, the orientation error approaches $\pi/2$, which destabilizes the system.

For $(K_f, K_\omega) = (1, 1)$, expression (4.49) can be deduced from the theoretical study for the error on flux orientation when $\omega_r = 0$. Expression (4.49) shows that when there is no error on R_s, but there is one on L_s, when $\omega_r = \omega_m = p\Omega = 0$, the orientation error approaches $\pi/2$, which destabilizes the system.

Combining cases $(K_f, K_\omega) = (0, 1)$ and $(K_f, K_\omega) = (1, 1)$ is especially interesting, since in the first case the sensitivity to uncertainties on inductance is low for a high mechanical speed, while in the second case this sensitivity is high for a low speed. Moreover, in both cases, the sensitivity to rotor resistance is null under steady conditions.

For $(K_f, K_\omega) = (1, 0)$, the theoretical study shows that the control sensitivity to all parameters and inductances is very high. Obviously, this case is of no interest.

5.3.3 Variations of the Consumption of Stator Current

Figure 5.17 shows the increase in the consumption of stator current by the machine in steady state.

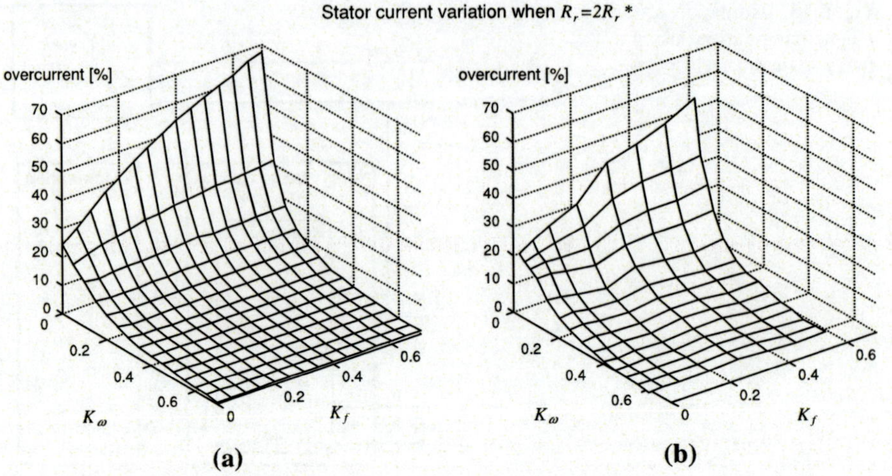

Fig. 5.17 Increase in the consumption of stator current depending on K_f and K when $R_r = 2R_r^*$ (speed = 1,500 rpm, torque = 2.3 Nm): **a** theoretical surface, **b** experimental surface

This increase results from a 100% error on the rotor resistance $(R_r = 2R_r^*)$, depending on K_f and K_ω, when the mechanical speed equals 1,500 rpm and the torque equals 2.3 Nm. The surface of Fig. 5.17b, as deducted from experimental measurements, confirms the theoretical predictions pictured by the surface of Fig. 5.17a. These figures show that the sensitivity of the system to errors on R_r depends mainly on K_ω and decreases strongly for low values of this coefficient. However, these diagrams show that when K_ω is close to 0, the coefficient K_f must not differ much from 0; otherwise the sensitivity increases strongly, which confirms that the case $(K_f, K_\omega) = (1, 0)$ is of no interest.

5.3.4 Introduction of a Fuzzy Logic Supervisor

5.3.4.1 Development of a Fuzzy Logic Supervisor

Values K_f and K_ω are chosen so as to reduce the flux control sensitivity to uncertainties on all electric parameters. As in the previous section, K_f and K_ω depend on mechanical speed ω_m and slip pulsation ω_r that reflects the torque. In the case studied here, the three stages of fuzzy logic enabling the determination of K_f and K_ω can be the following:

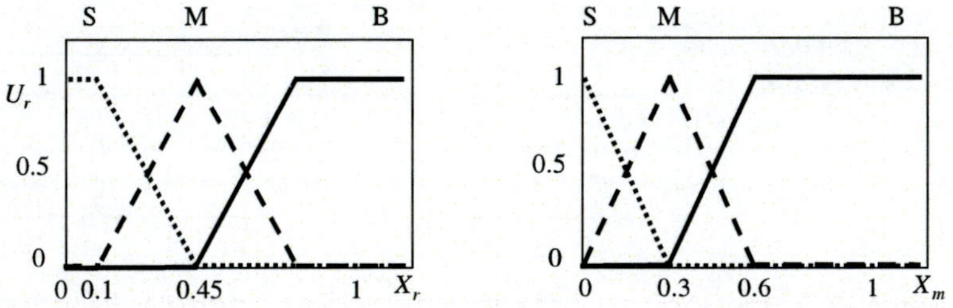

Fig. 5.18 Membership functions of the input variables

Table 5.6 States of output variables depending on the sensitivity to R_r

		X_r		
	(X_f, X_ω)	S	M	B
X_m	S	(X, X)	(X, B)	(X, B)
	M	(X, X)	(X, B)	(X, B)
	B	(S, X)	(X, B)	(X, B)

Fuzzification

Membership functions of normalized variables:

$$X_r = \left|\frac{\omega_r}{\hat{\omega}_r}\right| \quad X_m = \left|\frac{\Omega}{\hat{\Omega}}\right| \tag{5.12}$$

The membership functions are given in Fig. 5.18. In order to be able to define fuzzy rules, three states are assigned to these variables: big (B), medium (M) and small (S).

Inferences

The number of parameters taken into account in fuzzy logic is higher than in the case studied in the previous section (three states for each input variable, two output variables to determine). In order to determine fuzzy rules, information deduced from the theoretical study, simulations and experimental tests is carried forward in tables. Tables 5.6, 5.7 and 5.8 contain the states that should be assigned to output variables X_f and X_ω to reduce their sensitivity to R_r, R_s and M respectively. X stands for any state.

Table 5.9 contains information deduced from dynamic factors resulting from simulations and experimental tests.

Finally, Table 5.10 brings together the information contained in Tables 5.6, 5.7, 5.8 and 5.9. Some of the desired states are contradictory. In that case, the chosen value must correspond to the sensitivity that is to be reduced as a priority, i.e. the

Table 5.7 Table 5.6 States of output variables depending on the sensitivity to R_r

	(X_f, X_ω)	X_r S	M	B
X_m	S	(S, S)	(S, S)	(S, S)
	M	(S, X)	(X, X)	(X, X)
	B	(S, X)	(X, X)	(X, X)

Table 5.8 States of output variables depending on the sensitivity to M

	(X_f, X_ω)	X_r S	M	B
X_m	S	(X, S)	(B, S)	(X, S)
	M	(B, B)	(B, B)	(X, B)
	B	(B, B)	(B, B)	(X, B)

Table 5.9 States of output variables depending on the dynamic factors

	(X_f, X_ω)	X_r S	M	B
X_m	S	(X, X)	(X, X)	(X, X)
	M	(X, S)	(X, X)	(X, X)
	B	(S, S)	(S, X)	(S, X)

sensitivity to R_r in the case studied in this section. From Table 5.10 the following fuzzy rules can be deduced:

- if X_r is small and X_m is small, then X_f is small and X_ω is small, or
- if X_r is small and X_m is medium, then X_f is big and X_ω is small, or
- if X_r is small and X_m is big, then X_f is small and X_ω is small, or
- if X_r is medium and X_m is small, then X_f is small and X_ω is big, or
- if X_r is medium and X_m is medium, then X_f is big and X_ω is big, or
- if X_r is medium and X_m is big, then X_f is small and X_ω is big, or
- if X_r is big, then X_f is small and X_ω is big.

Defuzzification

The membership functions of output variables are given in Fig. 5.19. The result is as follows:

- $K_f = X_f$
- $K_\omega = X_\omega$ if $X_r X_m \geq 0$ else, $K_\omega = 0$ si $X_r X_m < 0$.

During the operating points corresponding to braking ($X_r X_m < 0$), k_ω is kept null so as to give the system an adequate dynamic.

Table 5.10 Synthesis of Tables 5.6, 5.7, 5.8 and 5.9

		X_r		
	(X_f, X_ω)	S	M	B
X_m	S	(S, S)	(S, B)	(S, B)
	M	(B, S)	(B, B)	(S, B)
	B	(S, S)	(S, B)	(S, B)

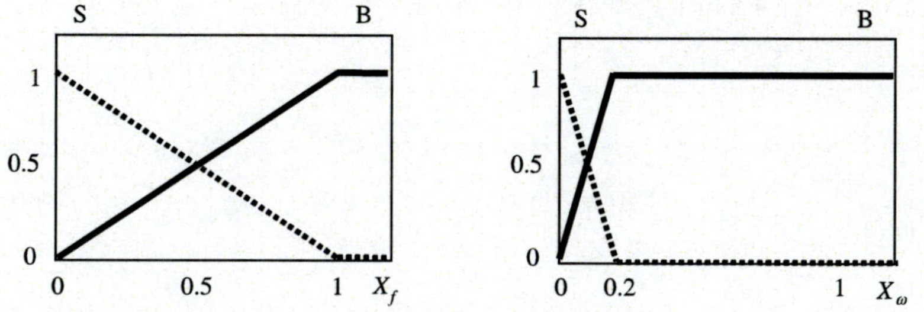

Fig. 5.19 Membership functions of the output variables

Figure 5.20 shows the evolution of K_f (5.20a) and K_ω (5.20b), deduced with fuzzy logic through the operating point of the machine.

5.3.4.2 Theoretical Comparison

To achieve the theoretical sensitivity analysis, coefficients A_1, B_1 and C in Eq. 4.5 can be determined as follows:

$$\begin{aligned} A_1 &= \left(1-K_f\right)A_1' + K_f A_1'' \\ B_1 &= \left(1-K_f\right)B_1' + K_f B_1'' \\ C &= \left(1-K_f\right)C' + K_f C'' \end{aligned} \tag{5.14}$$

Coefficients A_1', B_1' and C_1' are deduced from Eq 4.46, whereas A_1'', B_1'' and C'' are determined with the help of Eq 4.48. A_2, B_2 and D are deduced from Eq 5.8.

Figures 5.21, 5.22 and 5.23 show the sensitivity of the flux amplitude and orientation to errors on R_r, R_s and M respectively, depending on the mechanical speed and the electromagnetic torque. A 1% error is introduced on the parameter considered. These figures allow a comparison of the control sensitivity when $K_f = K_\omega$ (curves A) with the sensitivity when K_f and K_ω are determined through fuzzy logic (curves B).

Figure 5.21 clearly shows that a multi-model control allows for a dramatic drop in control sensitivity to uncertainties on the rotor resistance. The interest of the control strategy described in this section over the strategy described in the previous section is obvious when comparing Fig. 5.23b with Fig. 5.12b.

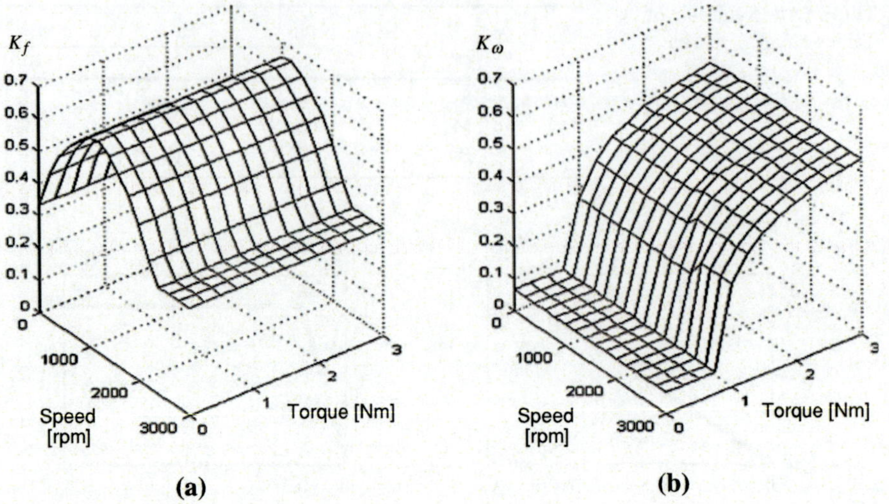

Fig. 5.20 Evolution of K_f (**a**) and K (**b**) as determined with fuzzy logic depending on the operating point of the machine

Considering only two estimates of the stator frequency result in a high sensitivity to uncertainties on inductances (5.2). The comparison of the two figures shows that this high sensitivity can be avoided through model double combination. Sensitivities obtained in Fig. 5.23b are overall even lower than sensitivities obtained through the standard control strategy, where $K_f = K_\omega = 0$ (Fig. 5.23a).

5.3.4.3 Experimental Comparison

Figures 5.24, 5.25 and 5.26 show the system's response to a reference speed step of 0–1,500 rpm, followed by a load torque step of 0–2.3 Nm (except in the case of Fig. 5.26b).

For each test, Figs. 5.24, 5.25 and 5.26 show the reference values and the measured value of mechanical speed, as well as the RMS value of stator current, respectively. These values are determined with optimized parameters, considering a 100% error on rotor resistance $\left(R_r = 2R_r^*\right)$ and a 20% error on inductance M ($M = 0.8\ M^*$).

Figure 5.24 also shows the evolution of coefficients K_ω and K_f. Each one of these figures describes three tests:

- Test A: $K_f = K_\omega = 0$;
- Test B: $K_f = 0$ and K_ω is determined through fuzzy logic (the same fuzzy logic presented in (5.2), but with an estimation of the magnetizing current);
- Test C: K_f and K_ω are determined through fuzzy logic.

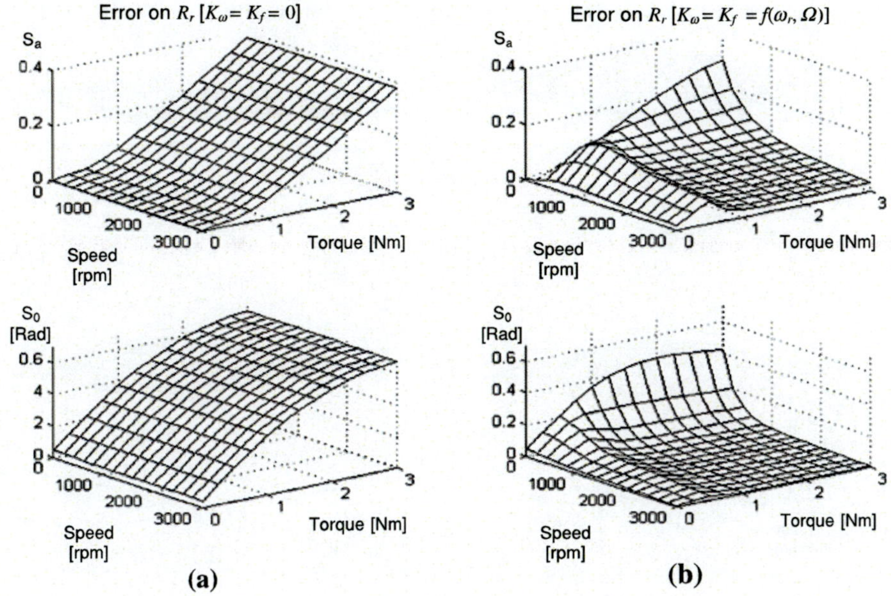

Fig. 5.21 Sensitivity of the flux amplitude and orientation to uncertainties on R_r, **a** $K_f = K = 0$, **b** K_f and K determined through fuzzy logic

The responses of speed and current are similar in the three tests compared in Fig. 5.24. These tests highlight the fact that the dynamic performances of the three control variations are almost identical when the parameters are optimized, despite the very fast variations of coefficients K_ω and K_f.

Comparing Fig. 5.25b, c with 5.25a clearly shows that multi-model controls allow for a dramatic drop in control sensitivity to uncertainties on R_r, as the speed response is faster and the stator current consumption lower in Fig. 5.25b, c.

The interest of the double combination control strategy becomes clear through the comparison of Figs. 5.26b and 5.26c.

The test in Fig. 5.26b corresponds to the single-combination control $(K_f = 0)$. In Sect. 4.2.2.2, we theoretically proved that for this control, in the event of errors on inductances, there is an instability risk when the machine is not loaded. The instability risk increases as mechanical speed increases, which is why the test in Fig. 5.26b is limited to 1,200 rpm. This test confirms the instability risk, even though no load torque is applied to the machine. Double combination eliminates the instability risk and the behavior of the system then becomes similar to the behavior of the standard control (Fig. 5.26c) as regards sensitivity to inductances is concerned.

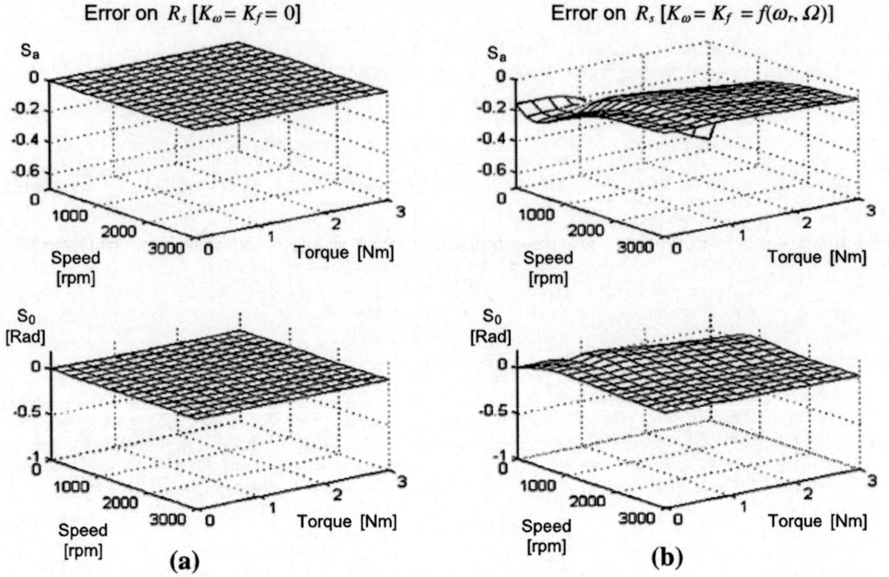

Fig. 5.22 Sensitivity of the flux amplitude and orientation to uncertainties on R_s, **a** $K_f = K = 0$, **b** K_f and K determined through fuzzy logic

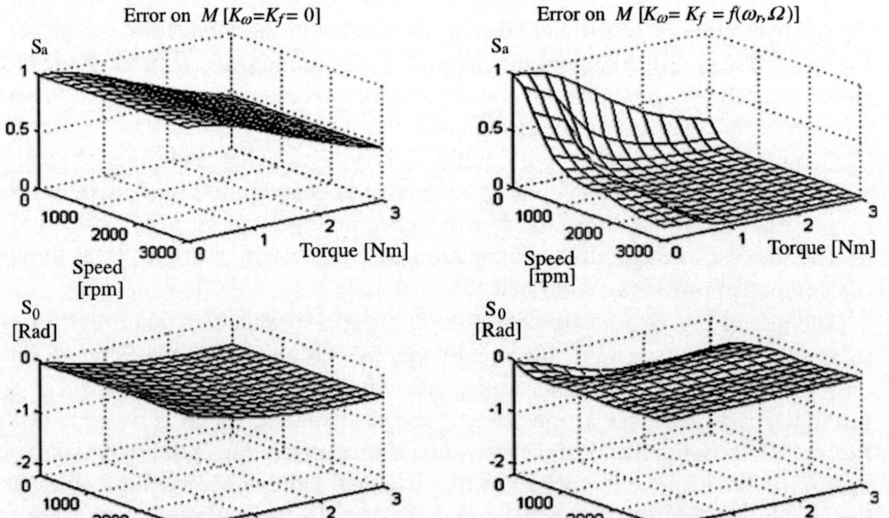

Fig. 5.23 Sensitivity of amplitude and flux orientation to uncertainties on M (not taking magnetic saturation into account); **a** $K_f = K = 0$, **b** K_f and K determined through fuzzy logic

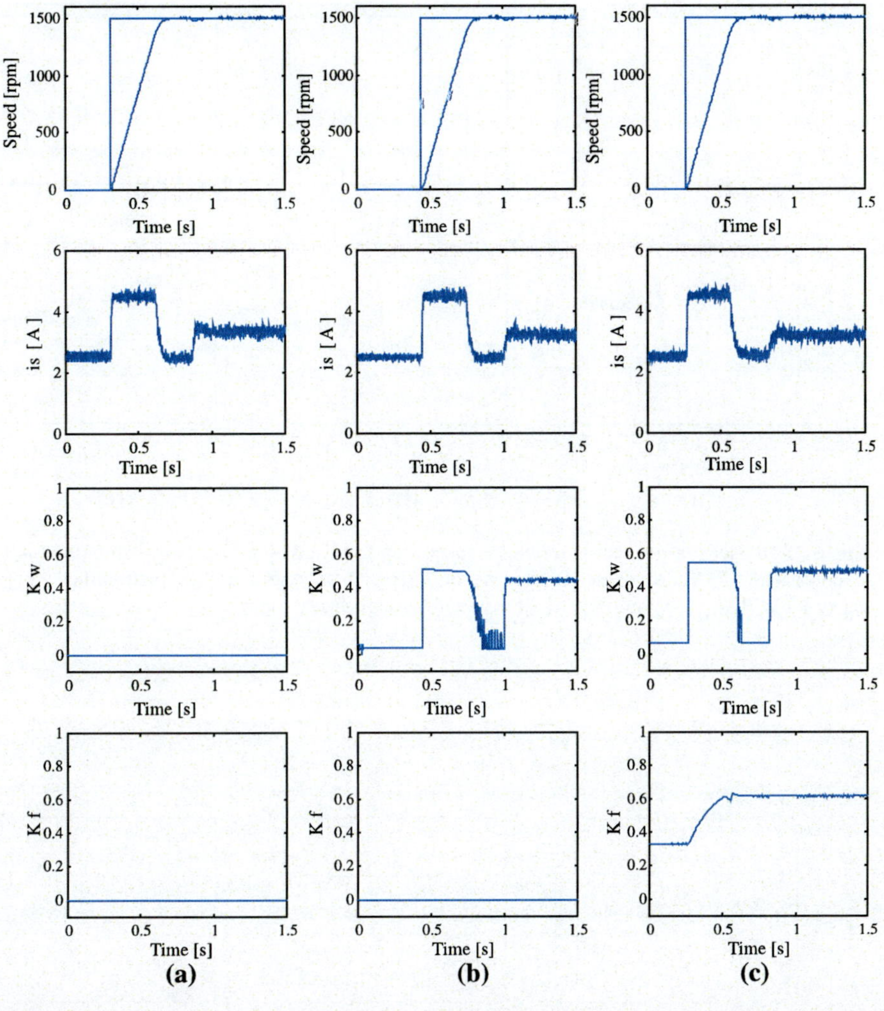

Fig. 5.24 System response to a reference speed step followed by a load torque step with optimized parameters. Test A: $K_f = K_\omega = 0$. Test B: $K_f = 0$ and K_ω is determined through fuzzy logic. Test C: K_f and K_ω are determined through fuzzy logic

5.4 Optimization of the Reduced-Order Rotor-Flux Observer

5.4.1 Introduction

An observer gain-choice method based on the study of parameter sensitivity and the optimization of different criteria was described in Chap.4. Fuzzy logic will enable us to integrate the information deduced from the sensitivity study,

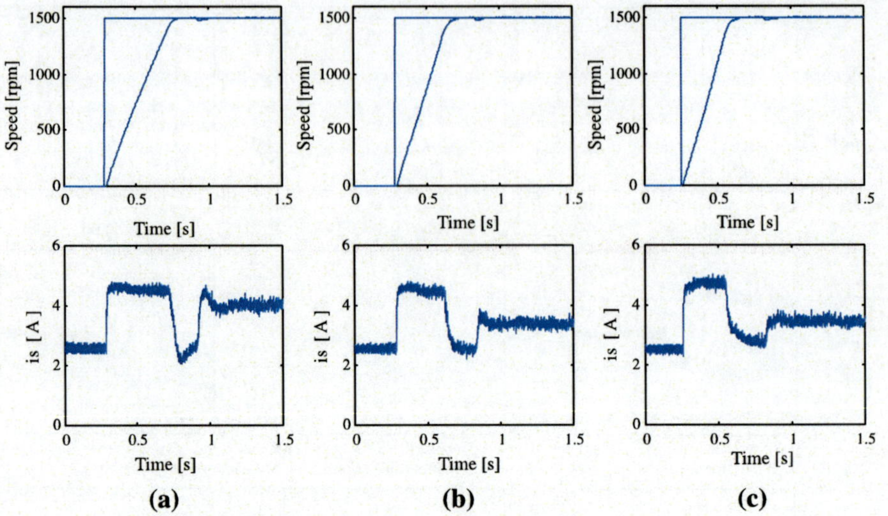

Fig. 5.25 System response to a reference speed step followed by a load torque step (tests A and C only) when $R_r = 2R_r^*$. Test A: $K_f = K_\omega = 0$. Test B: $K_f = 0$ and K_ω is determined through fuzzy logic. Test C: K_f and K_ω are determined through fuzzy logic

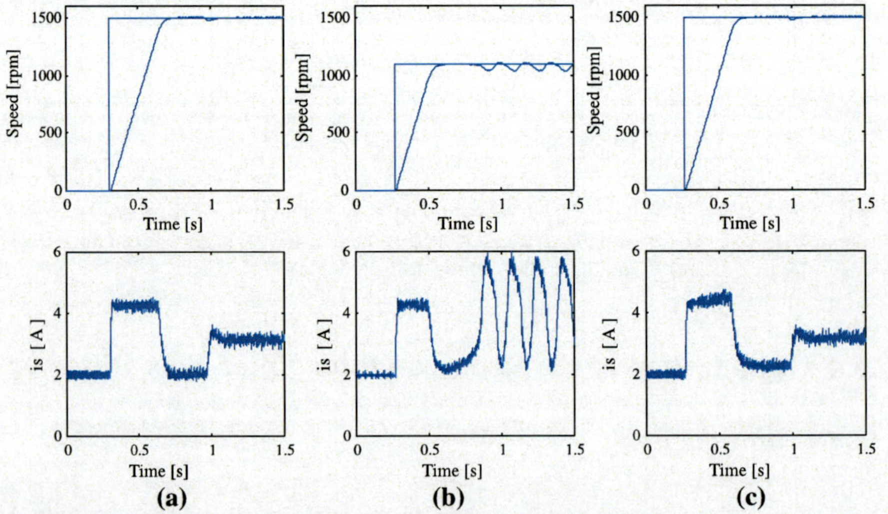

Fig. 5.26 System response to a reference speed step followed by a load torque step (tests A and C only) when $M = 0.8M^*$. Test A: $K_f = K_\omega = 0$. Test B: $K_f = 0$ and K_ω is determined through fuzzy logic. Test C: K_f and K_ω are determined through fuzzy logic

numerical simulations and experimental tests into the different optimization criteria described in Sect. 4.2.3.2.

Controls with reduced-order observers can be considered as multi-model controls since the observer is obtained by combining the rotor and stator equations of the machine.

In (Berthereau 2001b; Berthereau et al. 2002), the method introduced in this section is applied to a full-order observer.

5.4.2 *Sensitivity of the Reduced-Order Flux Estimator*

The rotor-flux estimator corresponds to the rotor-flux observer with null gains (k_1 and k_2, Eq. 3.80). Figure 5.27 shows amplitude and flux orientation errors, variations of the stator current, the evolution of the inductance M with the magnetic saturation, the real part of the poles and the relation between the imaginary part and the real part of those poles depending on the mechanical speed and the electromagnetic torque in the event of a 100% error on the rotor resistance $\left(R_r = 2R_r^*\right)$.

Errors committed on the flux, as well as the rise of the stator current consumption are then clearly noticeable. Note that the sensitivity obtained in this particular case is similar to the one obtained with the controls presented in the previous chapter in Fig. 4.6a, 5.10a and 5.21a, when the currents are controlled because of PI controllers and that $K_\omega = K_f = 0$ (which corresponds to the traditional versions of those controls). Figure 5.27 shows that the relation $\Im m/\Re e$ (pole) increases a lot when the speed increases, which results in an oscillating response from the observer.

5.4.3 *Determination of the Observer Gains with a Fuzzy Logic Supervisor*

5.4.3.1 Development of a Fuzzy Logic Supervisor

The fuzzy logic associated to the parametric sensitivity study will allow us to choose the observer gains k_1 and k_2 (Eq. 3.80) taking into account some criteria concerning:

- amplitude and flux orientation errors, (4.21) and (4.56);
- real and imaginary parts of the poles ($\Re e \leq (-R_r/L_r)$ and $\Im m/\Re e < 2$, for example);
- the overcurrent (4.4).

Those criteria must be checked for all operating points. k_1 and k_2 will then become functions of the measured speed $\widehat{\Omega}$ and of the estimated value of the slip pulsation ω_{r_reg} (which is proportional to the electromagnetic torque, Eq. 3.67).

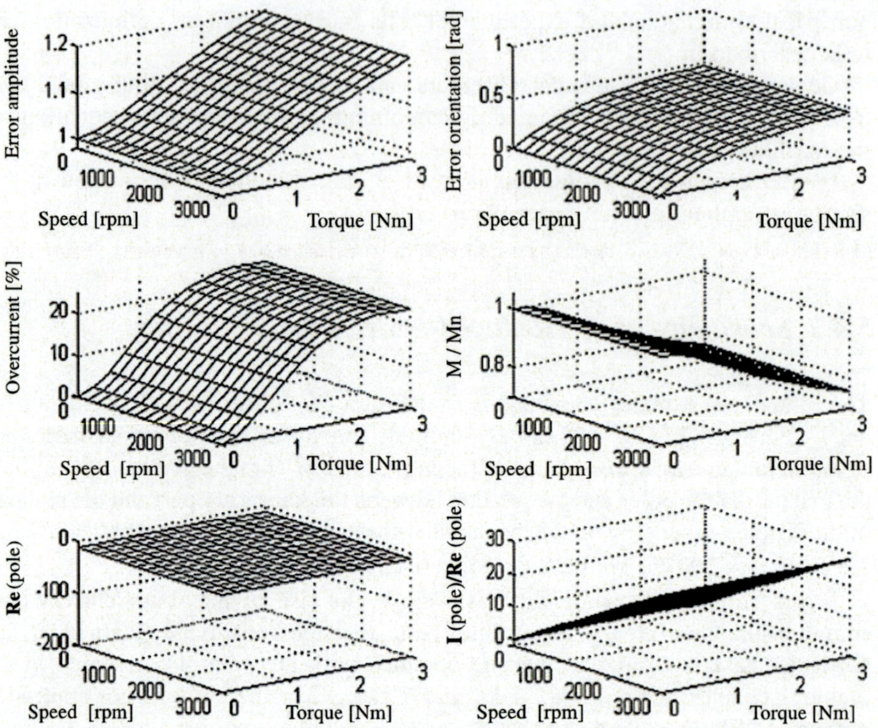

Fig. 5.27 Reduced-order flux observer developed in a reference frame linked to the stator. Theoretical results when $k_1 = k_2 = 0$ and $R_r = 2R_r$*

The following fuzzy algorithm was retained to determine some observer gains:

Fuzzification

Input variables are as follows:

$$X_r = \left|\frac{\omega_r}{\hat{\omega}_r}\right| \quad X_m = \left|\frac{\Omega}{\hat{\Omega}}\right| \tag{5.15}$$

In opposition to a priori controls in which K_ω and K_f are determined from the same membership functions of the input variables, k_1 and k_2 have different membership functions, when the input variables remain the same:

- The membership functions of the input variables for the determination of k_1 are presented in Fig. 5.28.
- The membership functions of the input variables for the determination of k_2 are presented in Fig. 5.29.

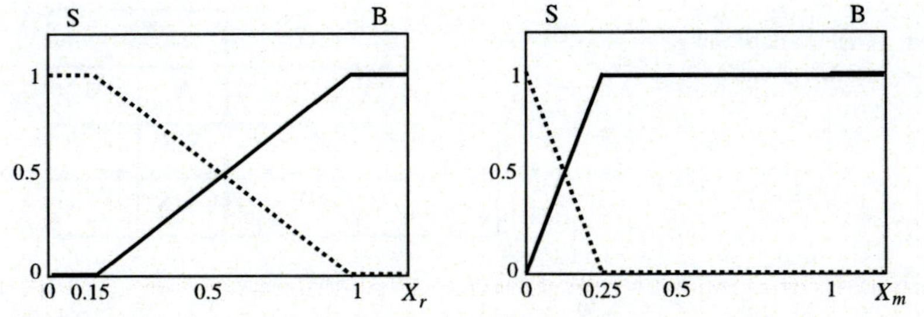

Fig. 5.28 Membership functions of the input variables for the determination of k_1

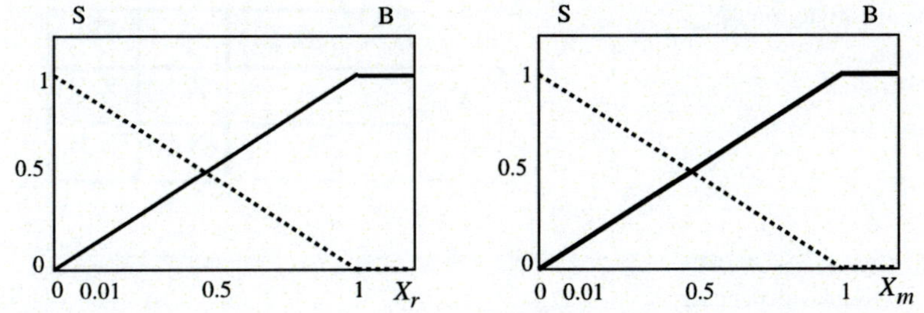

Fig. 5.29 Membership functions of the input variables for the determination of k_2

Table 5.11 States of the output variables depending on the sensitivity on R_r

		X_r	
	(X_{p1}, X_{p2})	S	B
X_m	S	(X,X)	(B,B)
	B	(X,X)	(X,S)

Inferences

X_{p1} and X_{p2} refer to output variables that will become k_1 and k_2 after defuzzification.

Inferences are set, in the same way as a priori controls, from the parametric sensitivity study applied to each parameter and from dynamic factors (Tables 5.11 5.12, 5.13 and 5.14).

The synthesis table is once again determined to primarily desensitize the command in comparison to R_r. The laws deducted from the synthesis Table 5.15 are the following:

- if X_r is small (S) or X_m is big (B) then X_{p1} is small (S), or
- if X_r is big (B) and X_m is small (S) then X_{p1} is big (B);
- if X_r is small (S) and X_m is small (S) then X_{p2} is small (S), or
- if X_r is big (B) or X_m is big (B) then X_{p2} is big (B).

Table 5.12 States of the output variables depending on the sensitivity on R_s

		X_r	
	(X_{p1}, X_{p2})	S	B
X_m	S	(S,S)	(S,S)
	B	(X,X)	(X,X)

Table 5.13 States of the output variables depending on the sensitivity on M

		X_r	
	(X_{p1}, X_{p2})	S	B
X_m	S	(X,X)	(X,X)
	B	(X,B)	(X,B)

Table 5.14 States of the output variables depending on the dynamic factors

		X_r	
	(X_{p1}, X_{p2})	S	B
X_m	S	(S,S)	(S,S)
	B	(X,B)	(X,B)

Table 5.15 Synthesis of Tables 5.11, 5.12, 5.13 and 5.14

		X_r	
	(X_{p1}, X_{p2})	S	B
X_m	S	(S,S)	(B,B)
	B	(S,B)	(S,B)

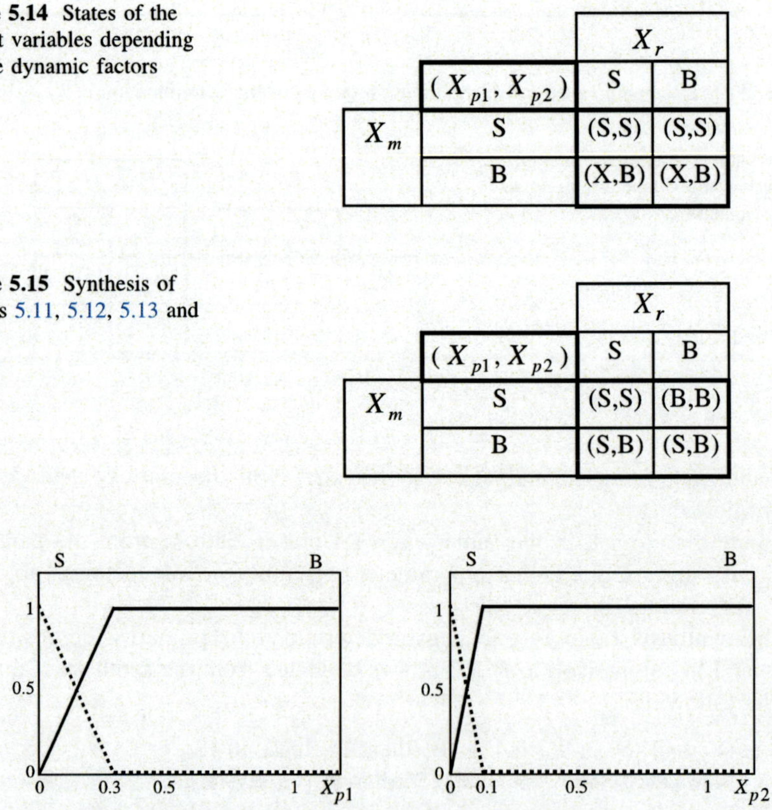

Fig. 5.30 Membership functions of output variables X_{p1} and X_{p2}

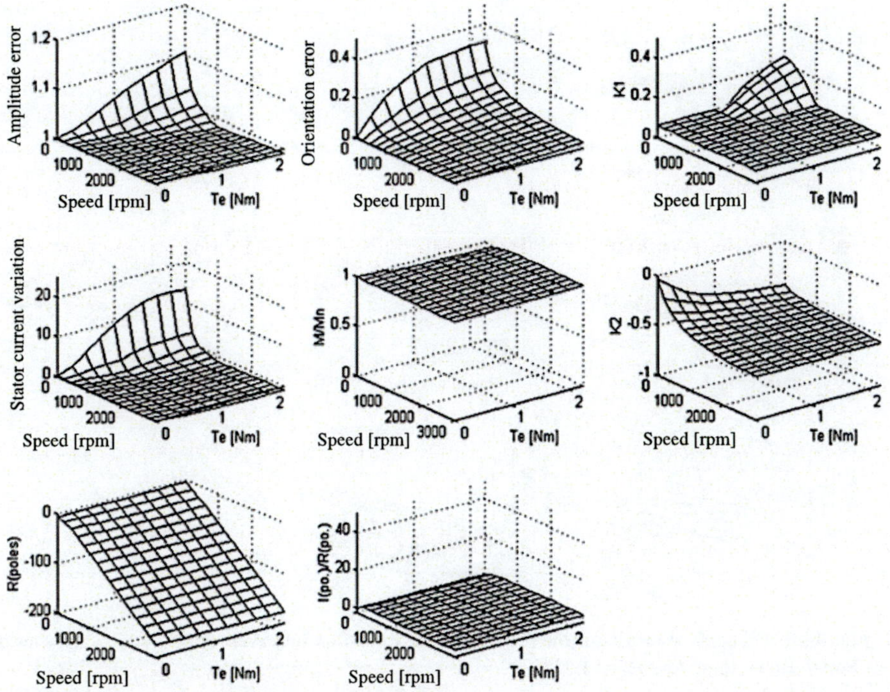

Fig. 5.31 Theoretical results of the reduced-order rotor-flux observer with k_1 and k_2 determined through fuzzy logic and $R_r = 2R_r^*$

Defuzzification

The membership functions of output variables X_{p1} and X_{p2} are represented in Fig. 5.30.

Other dynamic factors deducted from simulations and real-time tests added to stability criteria have shown that gains needed to be adjusted as follows:

$$\begin{aligned} &k_1 = 0.6X_{p1} \text{ and } k_2 = -1.1X_{p2} \text{ if } \Omega > 0 \text{ and} \\ &k_1 = 0.6X_{p1} \text{ and } k_2 = +1.1X_{p2} \text{ if } \Omega \leq 0 \end{aligned} \tag{5.16}$$

5.4.3.2 Theoretical Results

Figures 5.31, 5.32 and 5.33 represent flux errors, the stator current variation, the mutual inductance variation, the real part of the poles and the relation of the imaginary part on the real part of the poles as well as the values k_1 and k_2

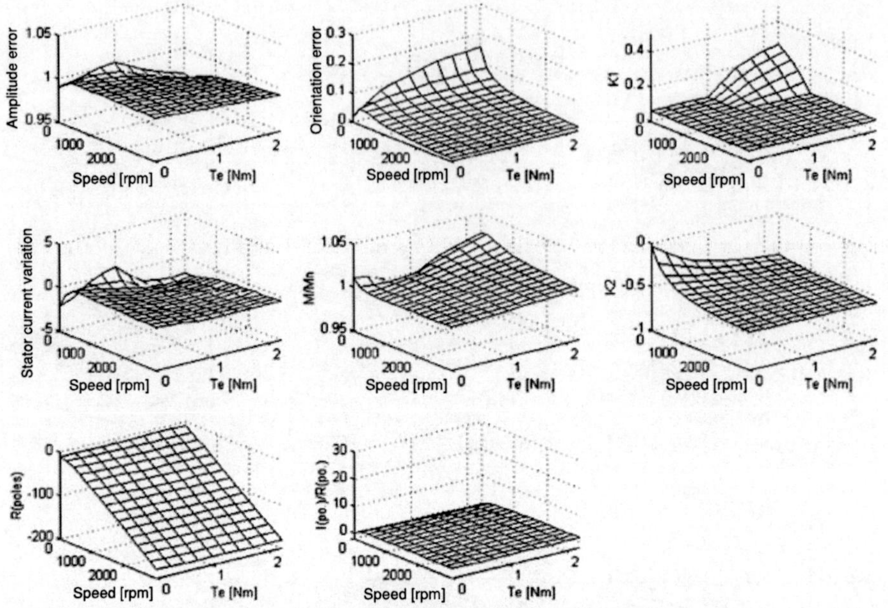

Fig. 5.32 Theoretical results of the reduced-order rotor-flux observer with k_1 and k_2 determined through fuzzy logic and $R_s = 1.2R_s^*$

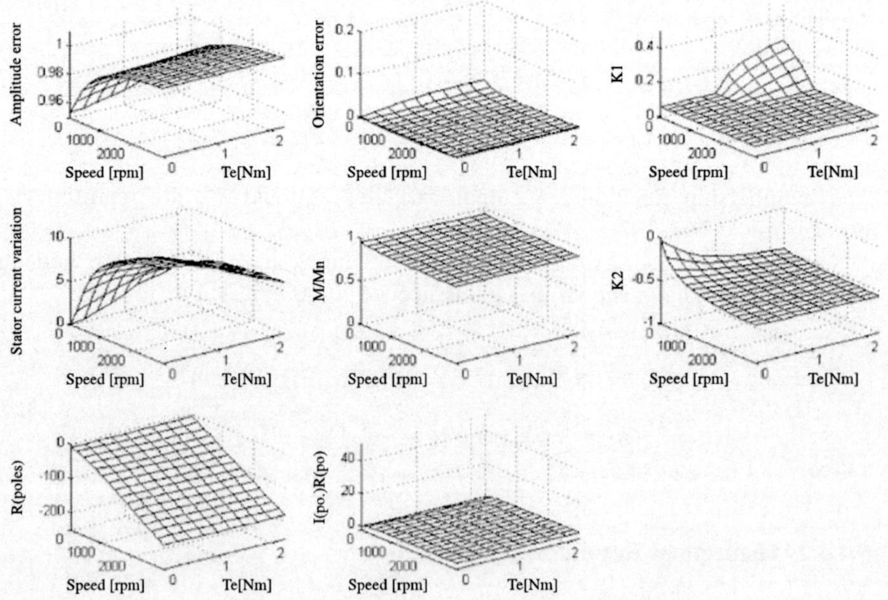

Fig. 5.33 Theoretical results of the reduced-order rotor-flux observer with k_1 and k_2 determined through fuzzy logic and $M = 0.8M^*$

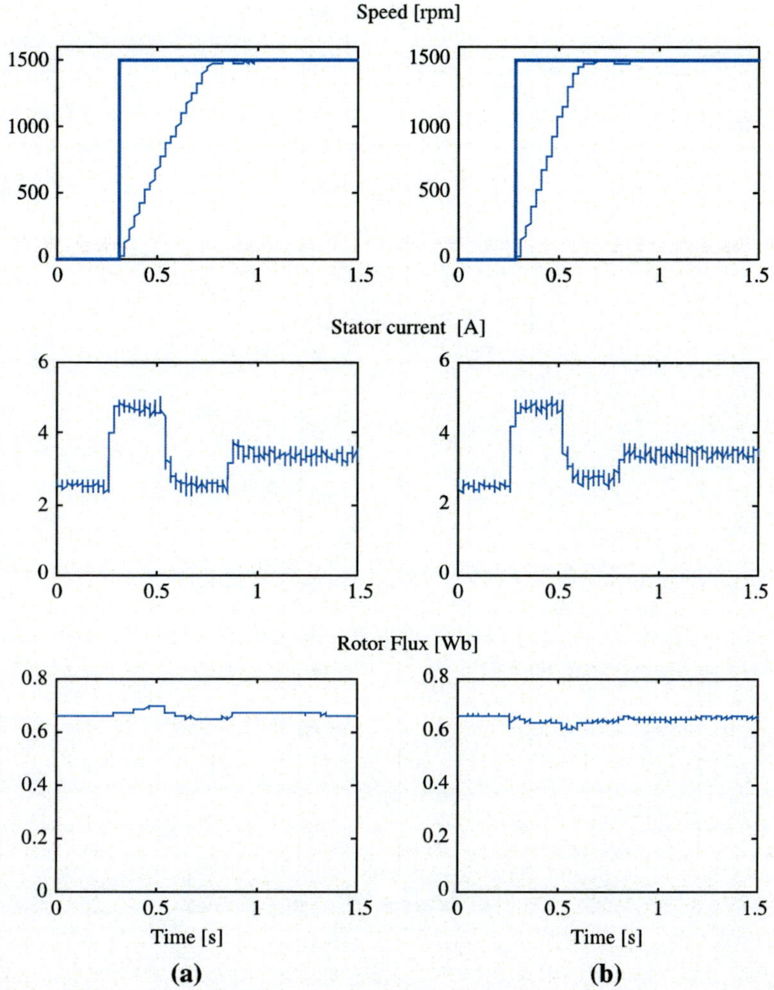

Fig. 5.34 Experimental response of the reduced-order rotor-flux observer at a step from 0 to 1,500 rpm followed by a torque step with optimized parameters, **a** $k_1 = k_2 = 0$, **b** k_1 and k_2 are determined through fuzzy logic

determined through fuzzy logic. Figure 5.31 is obtained by introducing a 100% error on the rotor resistance, a 20% error on the stator resistance for Fig. 5.32 and a 20% error on the mutual inductance for Fig. 5.33.

Figures 5.31, 5.32 and 5.33 show that the fuzzy logic allows to reach a satisfying compromise as regards the sensitivity to the different electric parameters. Poles vary depending on the operating point, but keep satisfactory values. Note that this interesting result has been obtained by the variation of the gains depending on the mechanical speed and the electromagnetic torque. With the

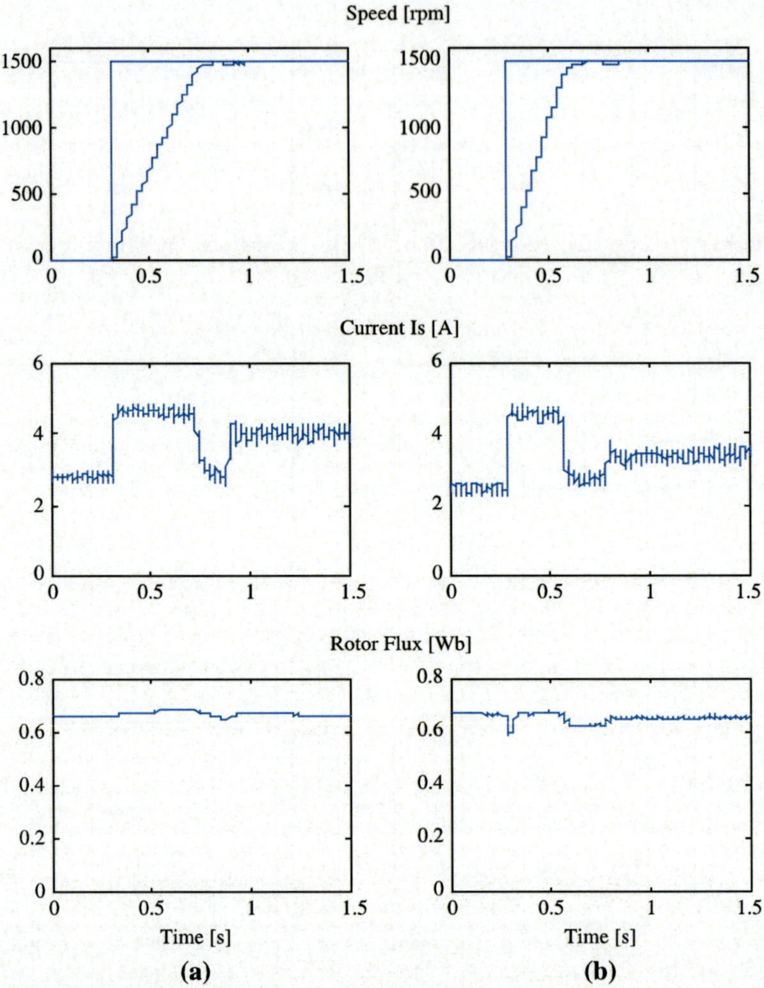

Fig. 5.35 Experimental response of the reduced-order rotor-flux observer at a step from 0 to 1,500 rpm followed by a torque step with a 100% error on the rotor resistance, **a** $k_1 = k_2 = 0$, **b** k_1 and k_2 are determined through fuzzy logic

classical method through pole placement, the gains vary only with mechanical speed.

5.4.3.3 Experimental Results

Figure 5.34 shows the speed of reference and the experimental response, the RMS stator current I_s and the response of the estimated flux at a reference step from 0 to 1,500 rpm followed by a torque step (2.3 Nm) with optimized parameters.

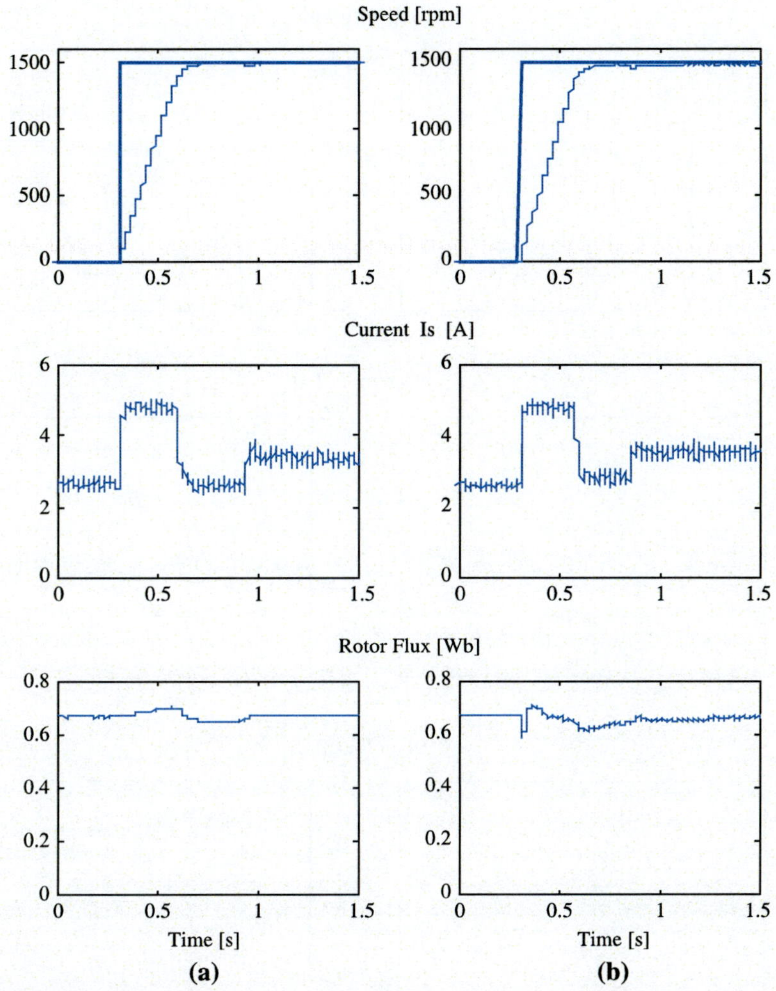

Fig. 5.36 Experimental response of the reduced-order rotor-flux observer at a step from 0 to 1,500 rpm with a 20% error on the stator resistance, **a** $k_1 = k_2 = 0$, **b** k_1 and k_2 are determined through fuzzy logic

For Fig. 5.34a, $k_1 = k_2 = 0$. For Fig. 5.34b, k_1 and k_2 are determined through fuzzy logic. The responses obtained with the two observer variations are identical.

Figure 5.35 shows the same tests as on Fig. 5.34 with a 100% error on the rotor resistance. The comparison between Figs. 5.35b and 5.35a shows the advantage of the reduced-order rotor-flux observer compared to the reduced-order rotor-flux estimator in the event of an error on the rotor resistance. Indeed, Fig. 5.35b shows:

- a decrease of the stator current consumption when the machine is loaded;
- an improvement of the speed establishment time thanks to a bigger available torque.

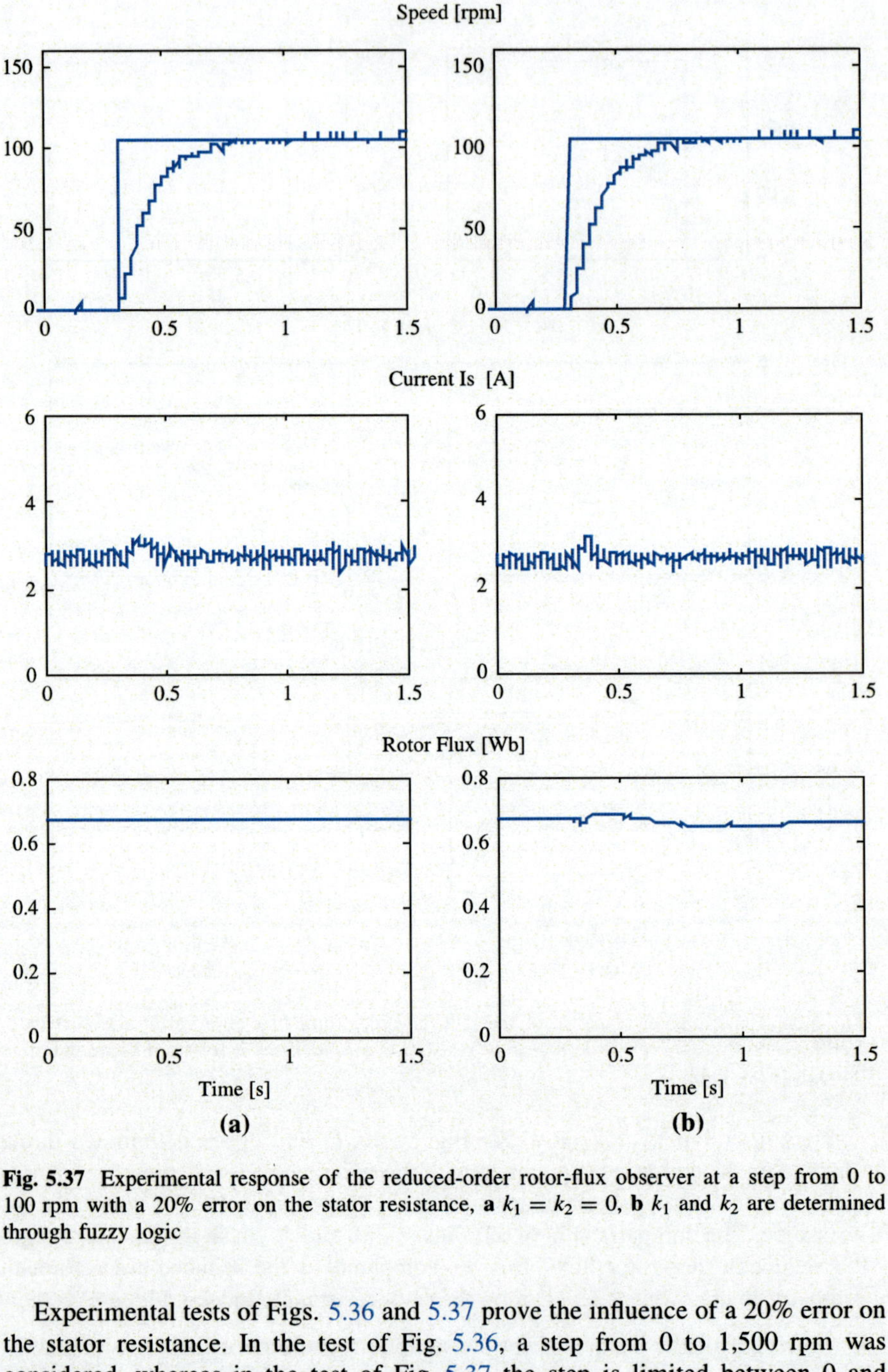

Fig. 5.37 Experimental response of the reduced-order rotor-flux observer at a step from 0 to 100 rpm with a 20% error on the stator resistance, **a** $k_1 = k_2 = 0$, **b** k_1 and k_2 are determined through fuzzy logic

Experimental tests of Figs. 5.36 and 5.37 prove the influence of a 20% error on the stator resistance. In the test of Fig. 5.36, a step from 0 to 1,500 rpm was considered, whereas in the test of Fig. 5.37 the step is limited between 0 and 100 rpm.

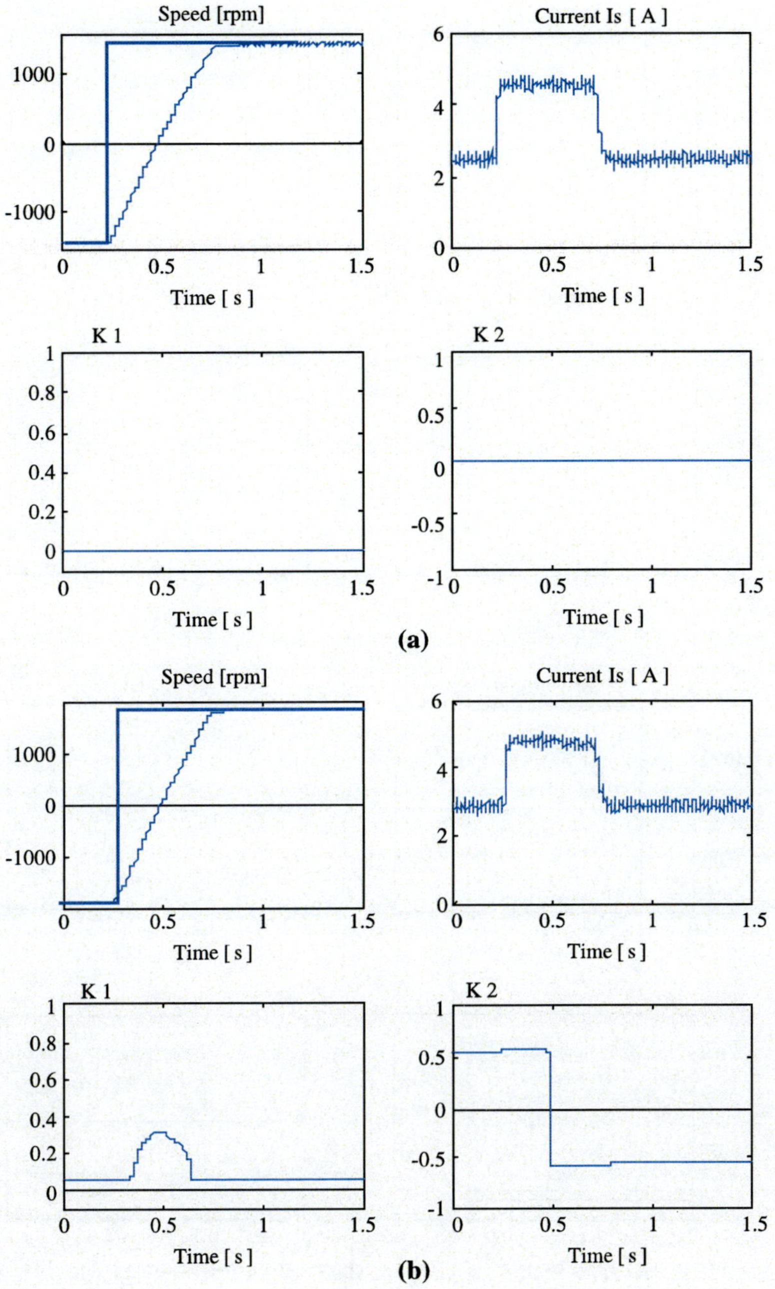

Fig. 5.38 Experimental response of the reduced-order rotor-flux observer at a step from −1,500 rpm to 1,500 rpm with optimized parameters, **a** $k_1 = k_2 = 0$, **b** k_1 and k_2 are determined through fuzzy logic

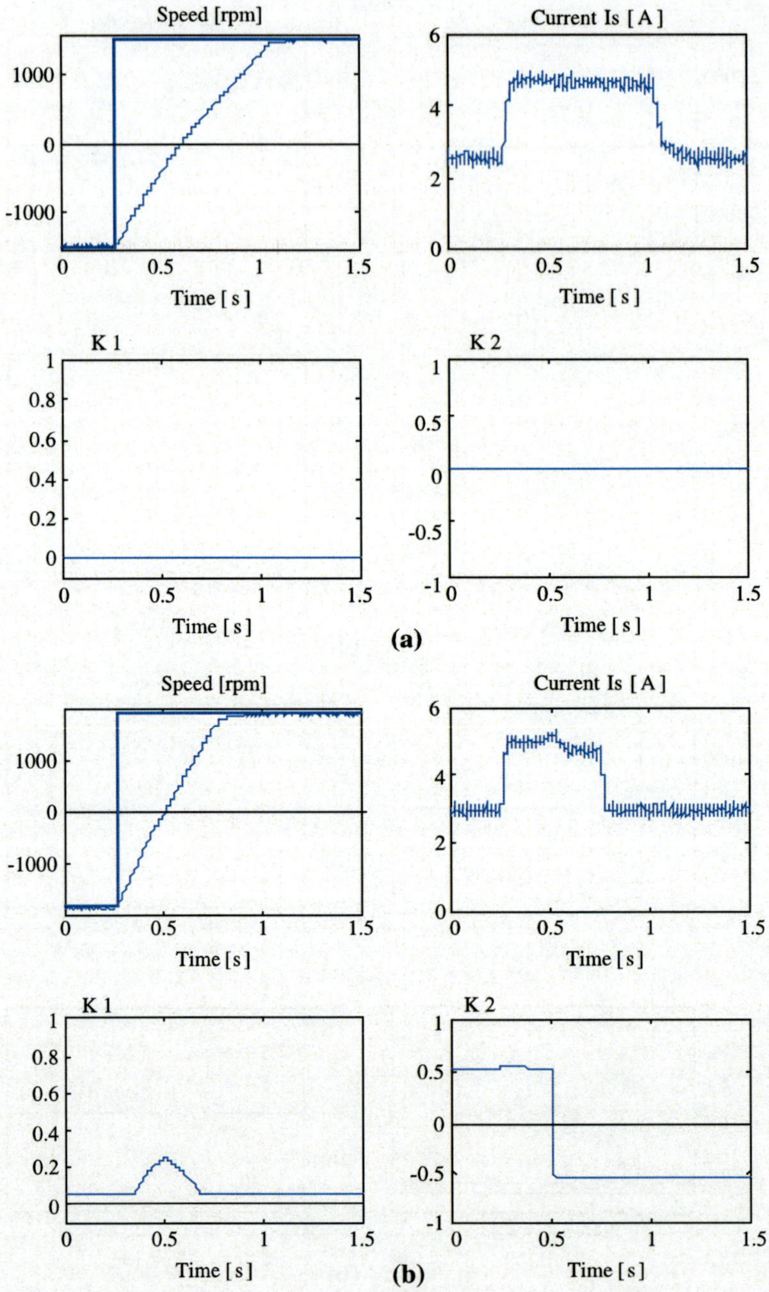

Fig. 5.39 Experimental response of the reduced-order rotor-flux observer at a step of 0–1,500 rpm with a 100% error on the rotor resistance, **a** $k_1 = k_2 = 0$, **b** k_1 and k_2 are determined through fuzzy logic

We can observe that the uncertainties on the stator resistance hardly influence the responses of Figs. 5.36b and 5.37b compared to those of Figs. 5.36a and 5.37a. Let's remind that the latter are obtained considering the flux estimator which is totally independent from the stator resistance.

Figure 5.38 represents the speed of reference and its experimental response, the RMS stator current I_s and the response of the gains k_1 and k_2 at a reference step of $-1{,}500$ to 1,500 rpm with optimized parameters. For Fig. 5.38a, $k_1 = k_2 = 0$. Figure 5.38b represents the results obtained with k_1 and k_2 determined through fuzzy logic. The results are similar considering the speed response. We can therefore notice a slight increase in the consumption of current when the gains are determined through fuzzy logic. This increase is due to the sampling period, which impacts the estimator and the flux observer differently (Delmotte et al. 2001).

Figure 5.39 reveals the same tests as shown in Fig. 5.38 introducing a 100% error on the estimated value of the rotor resistance. The comparison between Figs. 5.39a and 5.39b illustrates once again the capacities of the control based on the flux observer with fuzzy logic, for which we can notice a marked decrease of the speed settling time.

References

Bühler, H. (1994). *Réglage par logique floue*. Lausane: Presses polytechniques et universitaires romandes. ISBN: 2-88074-271-4.

Borne, P., Rozinoer, J., Dieulot, J. Y., Dubois, L. (1998). Introduction à la commande floue. Editions Technip. ISBN 2-7108-0721-1.

Robyns, B., Buyse, H., & Labrique, F. (1998). Fuzzy logic based field orientation in an indirect FOC strategy of an induction actuator. *Mathematics and Computer in Simulation, 46*, 265–274.

Robyns, B., Berthereau, F., Hautier, J. P., & Buyse, H. (2000). A fuzzy logic based multimodel field orientation in an indirect F.O.C. of an induction motor. *IEEE Transactions on Industrial Electronics, 47*(2), 380–388. (avril 2000).

Robyns, B., Labrique, F., & Buyse, H. (1997). Direct FOC of induction actuator using fuzzy logic to ensure field orientation with consideration of magnetic saturation effects. *Electromotion, 4*(1–2), 21–27.

Berthereau, F., Robyns, B., Hautier, J. P., & Buyse, H. (2001). Commande vectorielle multimodèle de la machine asynchrone avec superviseur à logique floue. *Revue Internationale de Génie Electrique, Hermès, 4*(3–4), 343–365.

Berthereau, F. (2001b). Commande vectorielle multialgorithmique appliquée à la machine asynchrone avec optimisation par supervision floue. PhD Thesus report from Université des Sciences et Technologies de Lille.

Berthereau, F., Robyns, B., & Hautier, J. P. (2002). Fuzzy logic based gain determination of induction motor flux observers. *EPE Journal, 12*(2), 19–25.

Delmotte, E., Semail, B., Robyns, B., & Hautier, J. P. (2001). Flux observer for induction machine control. Part I: Sensitivity analysis as a function of sampling rate and parameters variations. Part II: Robust synthesis and experimental implementation. *The European Physical Journal, Applied Physics, 14*, 13–43.

Chapter 6
Applications

This chapter aims at illustrating the interest of the controls shown in the previous chapter thanks to two applications:

- The kinetic energy storage, which is associated with a multi-energetic generation system made up of a wind turbine and a diesel generator set. In this application, an induction machine drives a flywheel which carries out the smoothing of the power generated depending both on the wind changes and the load consumption. This machine can be adapted to any need and works either as a motor, or a generator. Note that it nearly always operates under transient conditions. The control of this machine by means of a multi-model control allows for the optimization of the storage system's dynamic.
- The electrical traction system. A torque control of an induction machine performed with a reduced-order flux observer optimized through fuzzy logic will be introduced. This control tries to minimize errors on the torque estimate due to parametric uncertainties, especially for the most sensitive operating point under traction: speed equal to zero and maximum value for the torque which corresponds to the start-up period.

6.1 Kinetic Energy Storage System Combined with a Power System that Associates a Wind Turbine and a Diesel Generator

6.1.1 Introduction

In isolated geographical sites, the power supply is often performed by generator sets. However, most of these sites have a considerable wind energy potential and it is then worth associating wind turbines to generating sets. Their use is more cost-effective as the primary source of energy (wind) is completely free.

B. Robyns et al., *Vector Control of Induction Machines*, Power Systems,
DOI: 10.1007/978-0-85729-901-7_6, © Springer-Verlag London 2012

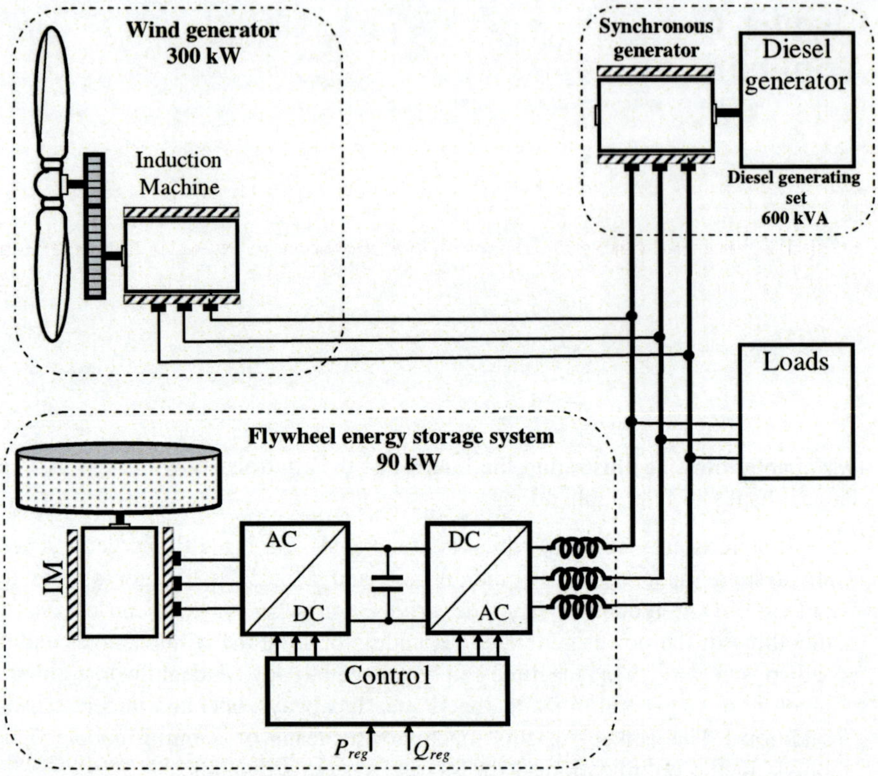

Fig. 6.1 Global diagram of the wind-diesel network including the kinetic energy storage system

Nevertheless, the power variations due to changes in the wind speed add to the power variations of the loads and lead to a faster wear of the diesel heat engine. An energy storage system can be used to absorb these variations and to reduce not only the fuel consumption but also the power variations the generating set is subjected to. A very dynamic short-term storage solution is based on a flywheel (Leclercq et al. 2003; Leclercq 2004; Cimuca 2005; Cimuca et al. 2006). Figure 6.1 represents the overall diagram of the power system discussed in this chapter. It is made up of a constant speed wind turbine of 300 kW, a generator set of 600 kVA, a collection of different loads that will remain constant—to simplify the study presentation—and a kinetic energy storage system. The system itself comprises an induction machine of 90 kW connected to the network thanks to a PWM double converter system and a flywheel 105 kg.m^2.

Since the stored energy depends on the square of the flywheel speed (with the relation $E = J\Omega^2/2$), the flywheel speed is kept between 3,000 and 6,000 rpm.

A fuzzy logic supervisor is suggested to reduce the power variations the generating set is subjected to and to control the power transfer between the network and the kinetic energy storage system. The vector control presented in Sect. 5.3—combining several algorithm variations of the flux control by means of a fuzzy logic supervisor—is used to control the induction machine coupled with the flywheel. This section points out that the lower parametric sensitivity of this control is particularly interesting in the considered application, which operates nearly always in dynamic state.

6.1.2 Control of the Kinetic Energy Storage System

6.1.2.1 Power Control Strategy

The kinetic energy storage system must compensate for the power variations generated by the wind turbine to reduce the power variations to which the generating set is subjected. If P_{reg} is the intended power from the association of the storage system and the wind turbine and P_{wg} is the power generated by the wind turbine, then the reference power transferred between the kinetic energy storage and the network is determined as follows:

$$P_{ref} = P_{reg} - P_{wg} \tag{6.1}$$

A filtered measure of the wind turbine power P_{wg} is used to determine P_{reg}, but since the flywheel speed is limited between 3,000 and 6,000 rpm, this speed needs to be taken into account to determine P_{reg} in order to avoid a saturation of the kinetic energy storage system. A fuzzy logic supervisor is then developed to determine P_{reg}.

6.1.2.2 Fuzzy Logic Supervisor of the Stored Power

Figure 6.2 points out the supervisor inputs that will lead to the determination of P_{reg}.

A. Fuzzification

The membership function of the input variable is represented in Fig. 6.3. Three fuzzy states are observed: small (S), medium (M) and big (B).

B. Inference

Since we aim at obtaining the most constant generated power, we will develop an expertise on the ability to impact the storage system using the universe of discourse previously determined for fuzzy rules. For example, if the wind power generated (P_{wg}) is big and if the flywheel speed is small, then the storage power is considerably big. The whole experience leads to the determination of nine fuzzy rules (Table 6.1). In order to refine the determination of the adjustment power, seven fuzzy states are considered for the output variable: very small (VS), small (S), small medium (SM), medium (M), big medium (BM), big (B) and very big (VB).

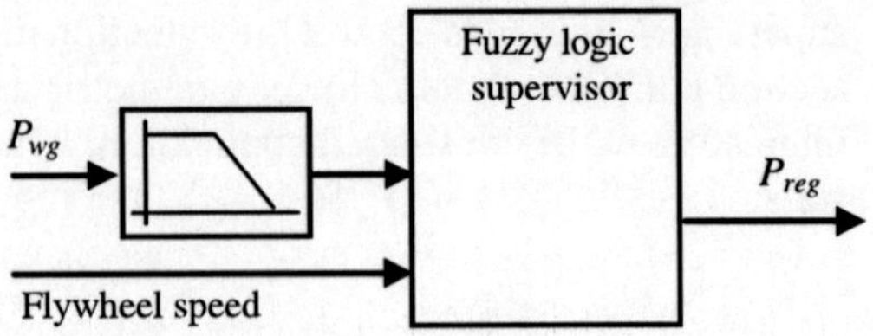

Fig. 6.2 Fuzzy logic supervisor

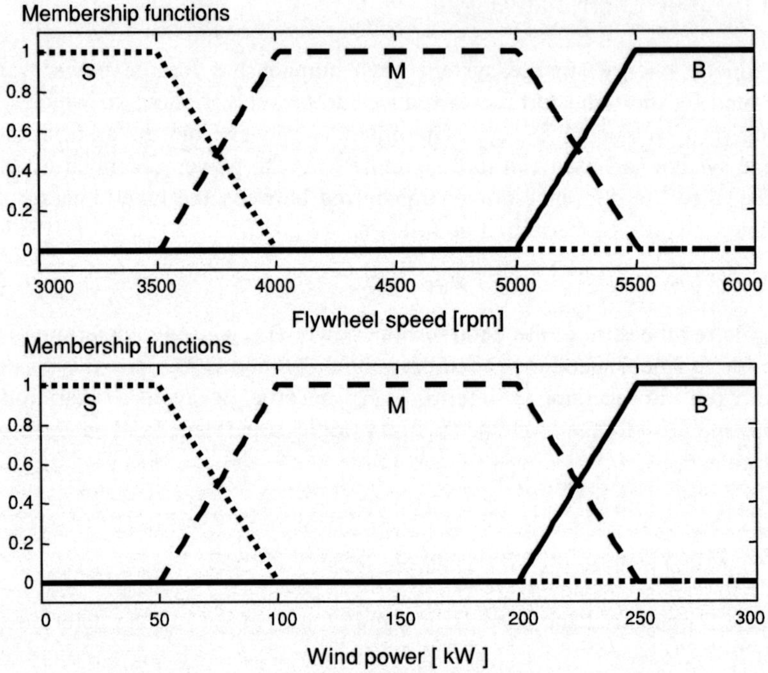

Fig. 6.3 Membership functions of the input variables

Table 6.1 Inference table

		P_{wg}		
	P_{reg}	Small (S)	Medium (M)	Big (B)
Flywheel speed	Small (S)	VS	SM	BM
	Medium(M)	S	M	B
	Big (B)	SM	BM	VB

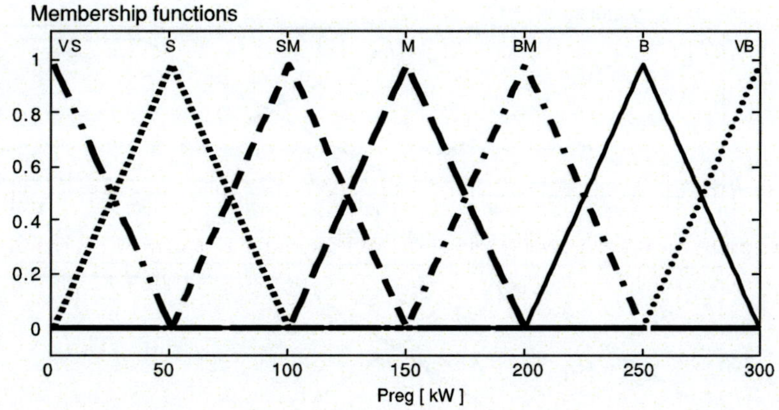

Fig. 6.4 Membership functions of the output variable

C. Defuzzification

The membership function of the output variable is represented in Fig. 6.4.

6.1.2.3 Supervision of the Induction Machine

The control strategy presented in Sect. 5.3 is applied to the induction machine driving the flywheel.

The induction machine mainly operates in overspeed (that is to say beyond its rated speed) and hence under defluxing conditions. If the flux is reduced, (3.67) shows that the slip pulsation is particularly high, with a constant current i_{sq} (a priori maximum in the considered application). Taking into account these observations, we will try to determine the coefficient values K_ω and K_f from the inputs: the rotor pulsation and the speed with a fuzzy algorithm (paragraph 5.3.4.1).

A. Fuzzification

Three fuzzy sets are observed: small (S), medium (M) and big (B). The membership function of the input variable is represented in Fig. 6.5.

B. Inference

Proceeding from these two fuzzy input quantities, the expertise leads to the determination of fuzzy values for the two coefficients. The fuzzy rules are delivered in Table 6.2. Two fuzzy states are considered for output variables: small (S) and big (B).

C. Defuzzification

The membership function of the output variables K_ω and K_f are represented in Fig. 6.6.

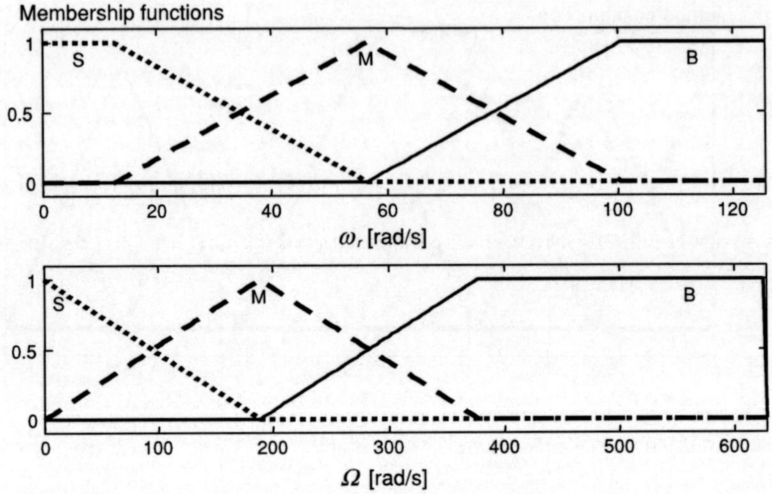

Fig. 6.5 Membership functions of the input variables of the multi-model vector control

Table 6.2 Inference table of the multi-model vector control

		ω_r		
	(K_f, K_ω)	S	M	B
Ω	S	(S,S)	(S,B)	(S,B)
	M	(B,S)	(B,B)	(S,B)
	B	(S,S)	(S,B)	(S,B)

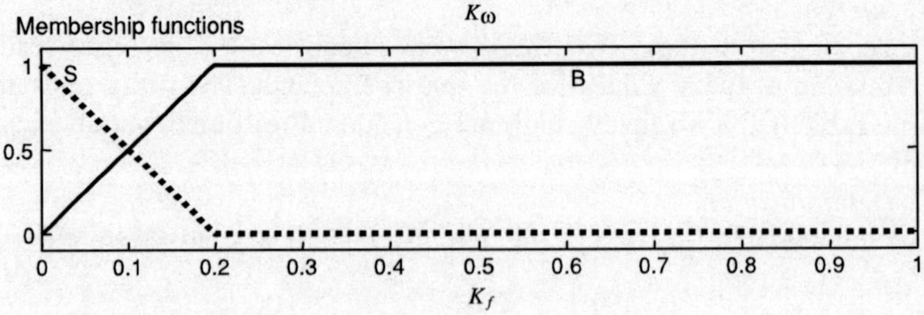

Fig. 6.6 Membership functions of the output variables of the multimodel control

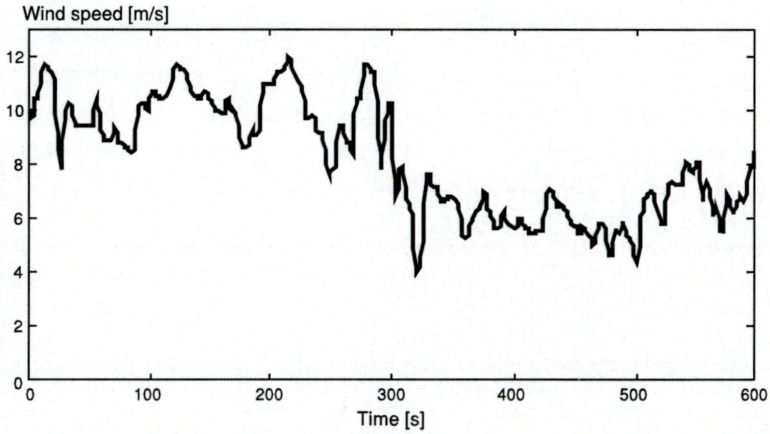

Fig. 6.7 Wind speed considered in the simulations

6.1.3 Results

Measures have been carried out on a constant speed wind turbine of 300 kW located in the North of France to simulate the wind turbine. Figure 6.7 shows the wind speed measured in the event of a medium wind speed (approximately 10 mps) followed by a low speed (about 6.3 mps). In the study, the active power needed by the load amounts to 300 kW and its reactive power reaches 120 kVar. At first, the rotation speed of the flywheel coupled with the induction machine amounted to 3,500 rpm.

The results shown in Figs. 6.8, 6.9, and 6.10 have been obtained considering the parameters of the induction machine were acknowledged and that $K_{\omega} = K_f = 0$, which corresponds to a standard vector control.

Figure 6.8 shows the active power generated by the generating set with and without kinetic energy storage system respectively in continuous lines or in dotted lines. This figure confirms that the suggested storage system allows for a considerable reduction of the power variations of the generating set. During the sudden wind speed variations, important power variations can still be observed.

Figure 6.9 shows that these variations increase when the intended power P_{reg} (indicated with dotted lines) cannot be followed by the considered system for which the power generated by the coupling between the wind turbine and the storage system is represented in continuous lines. This can be attributed to the power of the induction machine being coupled to the flywheel, which is limited to 90 kW.

The rotation speed of the flywheel is shown in Fig. 6.10. It shows no speed saturation because this speed has been taken into account for the determination of the reference speed (Table 6.1).

Figures 6.11, 6.12 and 6.13 show the response of the system to a 100% error simulation on the value of the rotor resistance ($K_{\omega} = K_f = 0$). The comparison of

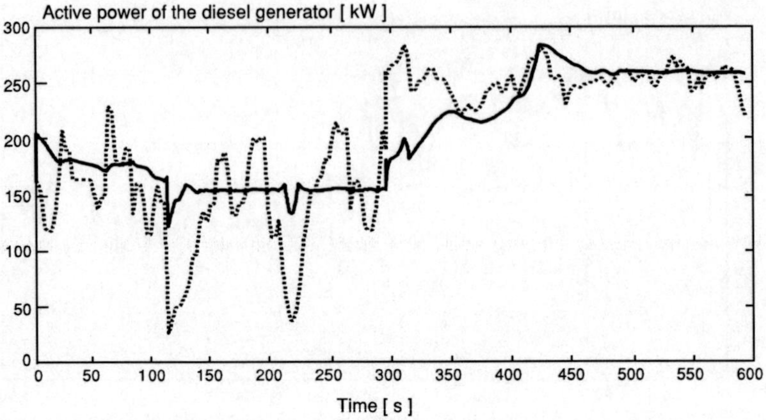

Fig. 6.8 Active power generated by the generating set with and without kinetic energy storage system, respectively in continuous lines and in dotted lines. $K_\omega = K_f = 0$, $R_r = R_r^*$

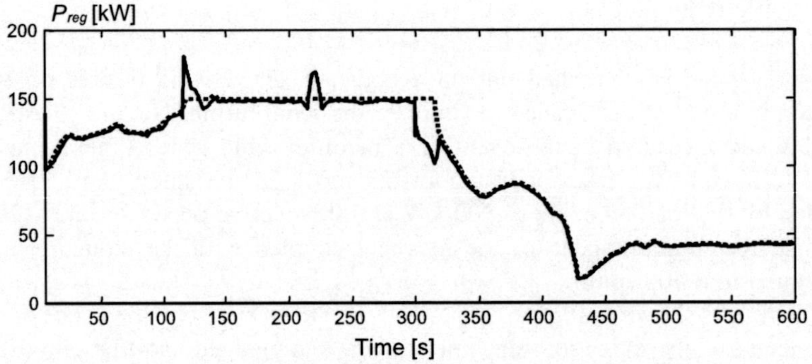

Fig. 6.9 Power generated by the combination of the kinetic storage and the wind turbine. Intended value P_{reg} in dotted lines and real value in continuous lines. $K_\omega = K_f = 0$, $R_r = R_r^*$

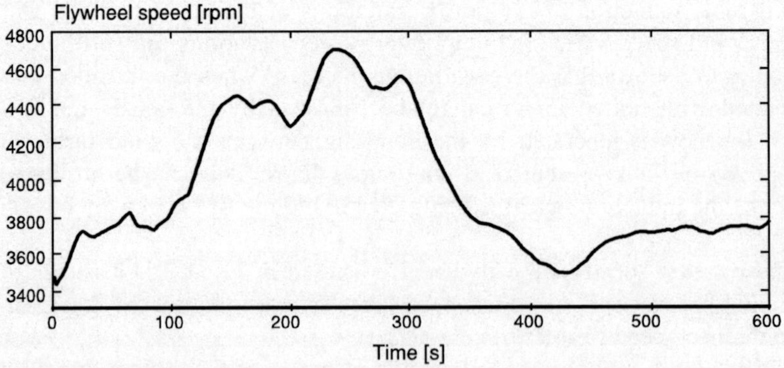

Fig. 6.10 Flywheel speed $K_\omega = K_f = 0$, $R_r = R_r^*$

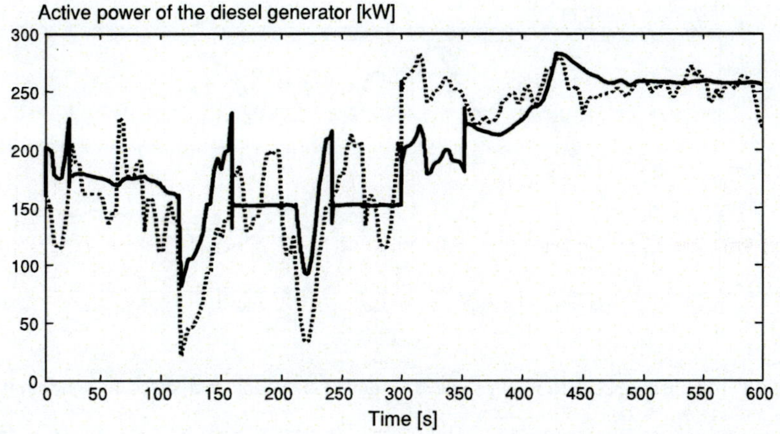

Fig. 6.11 Active power generated by the generating set with and without kinetic energy storage system, respectively in continuous lines and in dotted lines. $K_\omega = K_f = 0$, $R_r = 2R_r^*$

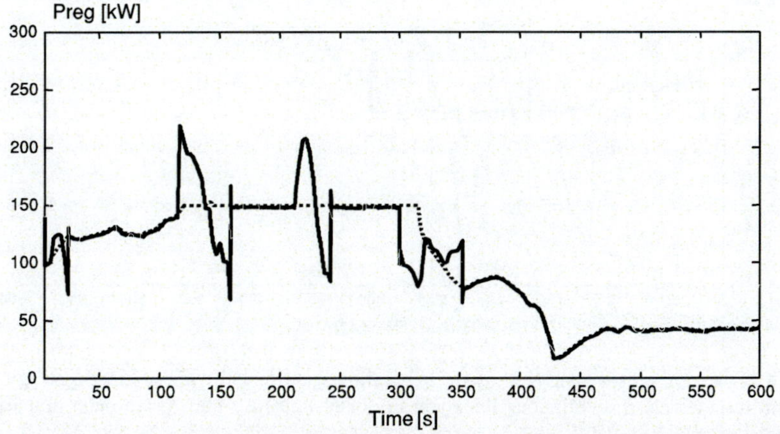

Fig. 6.12 Power generated by the combination between kinetic storage and wind turbine. Intended value P_{reg} in dotted lines and real value in continuous lines. $K_\omega = K_f = 0$, $R_r = 2R_r^*$

Fig. 6.11 with Fig. 6.8 shows that the power variations generated by the generating set—in which a storage system is used—remain considerably high.

Figure 6.12 proves that the power generated by the association of the storage system and the wind turbine is not proportional to the intended power P_{reg}, even with important changes in wind speed. This can be attributed from the difference between the real speed and the reference speed of the induction machine shown respectively in continuous lines and in dotted lines in Fig. 6.13.

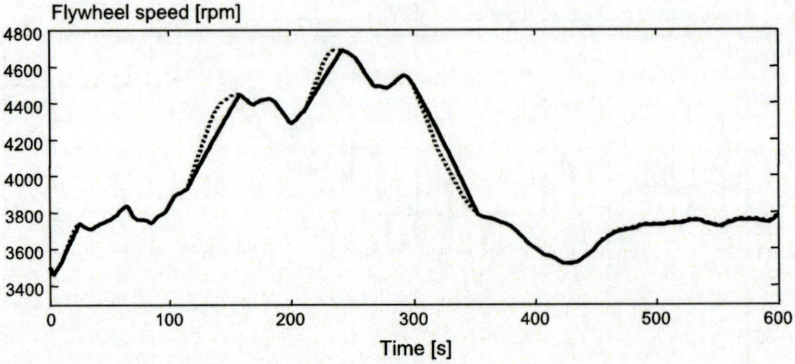

Fig. 6.13 Flywheel speed. Reference speed in dotted lines and real speed in continuous lines. $K_\omega = K_f = 0$, $R_r = 2R_r^*$

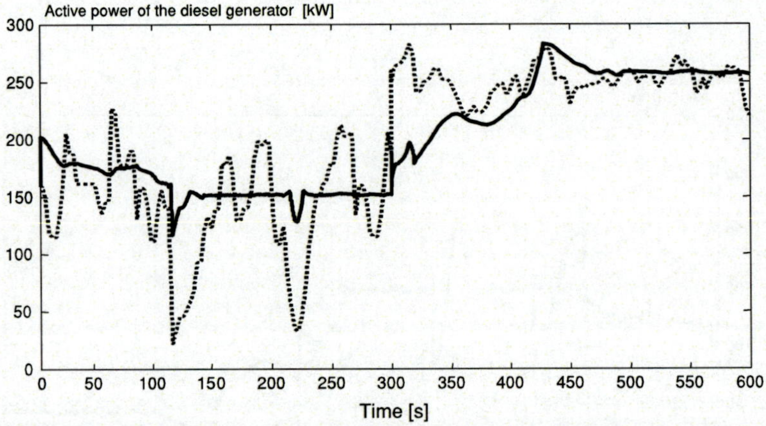

Fig. 6.14 Active power generated by the generating set with and without kinetic energy storage system respectively in continuous lines and in dotted lines. K_ω and K_f are determined through fuzzy logic, and $R_r = 2R_r^*$

This error is induced by the error on the rotor resistance, which leads to a reduction of the electromagnetic torque in the machine and prevents the development of the optimal torque.

Figures 6.14, 6.15 and 6.16 illustrate the system response when K_ω and K_f are determined by means of the fuzzy algorithm presented in 6.1.2.3, with a 100% error on the rotor resistance value. The results are similar to those presented in Figs. 6.8, 6.9, and 6.10 with a standard control without any errors on the parameters. These results confirm the interest of the multimodel control with fuzzy logic supervisor in the considered application.

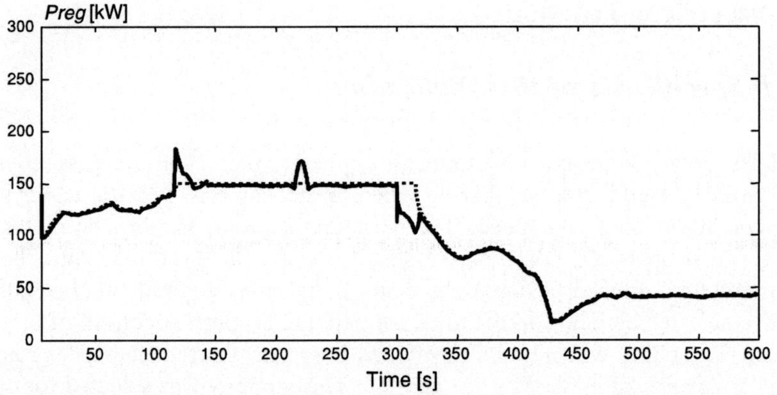

Fig. 6.15 Power generated by the association of kinetic storage and wind turbine. Intended value P_{reg} in dotted lines and real value in continuous lines K_{ω} and K_f are determined through fuzzy logic, and $R_r = 2R_r^*$

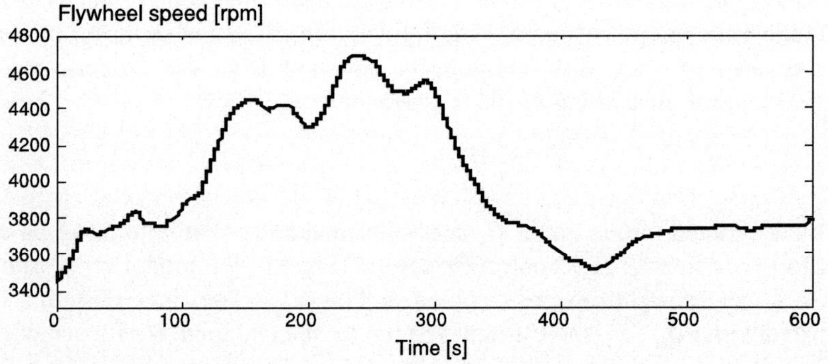

Fig. 6.16 Flywheel speed. Reference speed in dotted lines and real speed in continuous lines. K_{ω} and K_f are determined through fuzzy logic, and $R_r = 2R_r^*$

6.1.4 Parameters of the Induction Machine Coupled with the Flywheel

Rated power:	90 kW
Number of poles:	2
Stator resistance:	15.8 mΩ
Rotor resistance:	14.5 mΩ
Mutual inductance:	18.9 mH
Stator inductance:	19.2 mH
Rotor inductance:	19.3 mH

6.2 Electrical Traction

6.2.1 *Specificities of the Application*

Vehicles move when a machine torque is applied to them. Currently, speed control is sometimes used for easier driving when the machine is working under steady-state conditions (constant speed). For these applications, starting the motor is the most critical operating point, as the speed is null and the torque must be at its maximum. In our example, a traction chain is driven by an induction machine, like for instance an electrical forklift truck for pallets transport. A control of the torque of such a machine will now be developed based on the reduced-order rotor flux observer developed in Sects. 4.2.3 and 5.4. This strategy was selected for offering the best results for the null-speed critical operating point (Berthereau 2001).

Torque control, of paramount importance here, is realized through closed-loop control (Fig. 6.17). Measuring torque is both an expensive and a delicate operation. Therefore, the estimator has been developed in line with the methodology presented in paragraph 3.1.5.3. The machine's electromagnetic torque is estimated in a static Park reference frame. This frame is linked to the stator on the basis of the estimates of rotor flux components obtained from the reduced-order flux observer and the measuring of stator currents (Eq. 2.106):

$$\tilde{T} = p\frac{M^*}{L_r^*}\left(\tilde{\phi}_{r\alpha}\widehat{i}_{s\beta} - \tilde{\phi}_{r\beta}\widehat{i}_{s\alpha}\right) \tag{6.2}$$

This estimated torque and a C_T controller make it possible to design a closed-loop torque control. The control structure (Fig. 5.15), modified, is shown in Fig. 6.17. The control device is shown in Fig. 6.18. The rotor pulsation can be estimated with Eq. 3.67 and estimates of the torque and the rotor flux components.

6.2.2 *Study of Sensitivity on Torque Control*

To evaluate parameter sensitivity, the electromagnetic torque must be determined as a function of the reference torque and of the real and estimated machine parameters. Since the flux errors are defined in a reference frame linked to the rotating field, expression (6.2) must be transposed into this reference frame. The Park transform gives:

$$\tilde{\phi}_{r\alpha} = \tilde{\phi}_{rd}\cos\theta_s - \tilde{\phi}_{rq}\sin\theta_s \tag{6.3a}$$

$$\tilde{\phi}_{r\beta} = \tilde{\phi}_{rd}\sin\theta_s + \tilde{\phi}_{rq}\cos\theta_s \tag{6.3b}$$

Introducing the expressions (6.3a, b) in (6.2) results in:

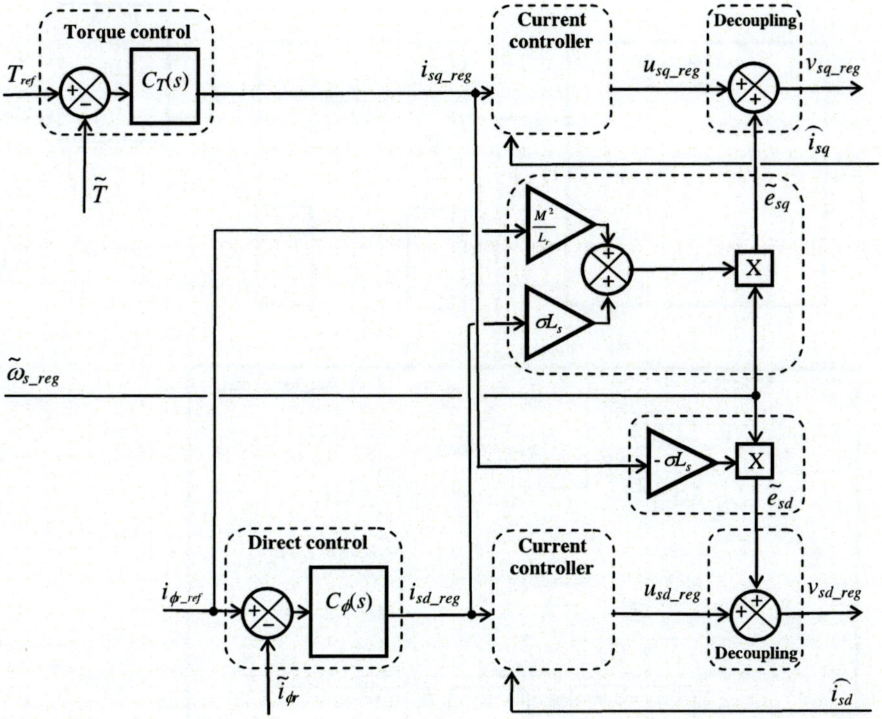

Fig. 6.17 Control diagram for the study of the rotor flux observer—traction is controlled through torque control

$$\tilde{T} = p\frac{M^*}{L_r^*}\left(\tilde{\phi}_{rd}\widehat{i}_{s\beta}\cos\theta_s - \tilde{\phi}_{rq}\widehat{i}_{s\beta}\sin\theta_s - \tilde{\phi}_{rd}\widehat{i}_{s\alpha}\sin\theta_s - \tilde{\phi}_{rq}\widehat{i}_{s\alpha}\cos\theta_s\right) \tag{6.4}$$

Considering that:

$$i_{sd} = i_{s\alpha}\cos\theta_s + i_{s\beta}\sin\theta_s \tag{6.5a}$$

$$i_{sq} = -i_{s\alpha}\sin\theta_s + i_{s\beta}\cos\theta_s \tag{6.5b}$$

Using (6.5a, b) and (6.4), the torque estimation can be defined as:

$$\tilde{T} = p\frac{M^*}{L_r^*}\left(\tilde{\phi}_{rd}\widehat{i}_{sq} - \tilde{\phi}_{rq}\widehat{i}_{sd}\right) \tag{6.6}$$

To evaluate parameter sensitivity, the electromagnetic torque must be determined as a function of the reference torque and the real and estimated parameters of the machine. Since the flux controller includes an integral action, it is admitted that:

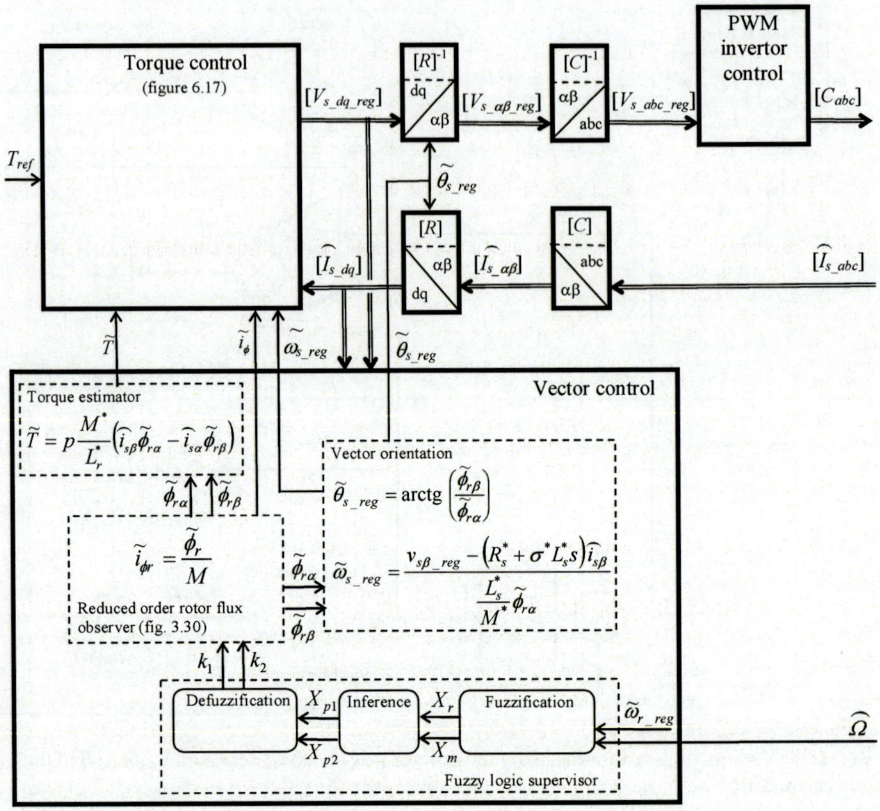

Fig. 6.18 Control diagram for the study of the rotor flux observer—traction is controlled through torque control

$$\left|\tilde{\phi}_r\right| = \phi_{r\_\text{ref}} \tag{6.7}$$

Working from Fig. 4.2 to design a vector control of rotor flux and keeping in mind that only the estimated values are controlled and oriented, the equations are:

$$\tilde{\phi}_{rd} = \phi_{r\_\text{ref}}\cos(\rho_o + \rho_e) = \phi_{r\_\text{ref}} \tag{6.8a}$$

$$\tilde{\phi}_{rq} = \phi_{r\_\text{ref}}\sin(\rho_o + \rho_e) = 0 \tag{6.8b}$$

With these expressions, the torque estimate becomes:

$$\tilde{T} = p\frac{M^*}{L_r^*}\phi_{r\_\text{ref}}\hat{i}_{sq} \tag{6.9}$$

The stator current on axis q remains to be determined. Since the sensitivity study was realized in steady state, (4.25d) means that:

$$i_{sq} = \frac{L_r}{M\,R_r}\omega_r\,\phi_{rd} + \frac{1}{M}\phi_{rq} \tag{6.10}$$

The rotor flux components are determined with the help of Fig. 4.2:

$$\phi_{rd} = \frac{|\phi_r|}{\phi_{r\_ref}}\phi_{r\_ref}\cos\rho_o \text{and} \phi_{rq} = \frac{|\phi_r|}{\phi_{r\_ref}}\phi_{r\_ref}\sin\rho_o \tag{6.11}$$

By introducing (6.11) in (6.10), and (6.10) in (6.9) and using (4.22):

$$\tilde{T} = p\frac{M^*}{L_r^*}\frac{\phi_{r\_ref}^2}{\sqrt{q_1^2+q_2^2}}\left(\omega_r\frac{L_r}{M\,R_r}\cos\rho_o + \frac{1}{M}\sin\rho_o\right) \tag{6.12}$$

The following expression can also be deduced from Fig. 4.2:

$$\underline{\phi_r} = (\cos\rho_o + j\sin\rho_o)\,\phi_{r\_ref} \tag{6.13}$$

According to expression (4.18):

$$\underline{\tilde{\phi}_r} = (q_1 + jq_2)\,\underline{\phi_r} \tag{6.14}$$

Identifying the above equations results in:

$$\underline{\phi_r} = \frac{\tilde{\phi}_r}{q_1 + jq_2} = (\cos\rho_o + j\sin\rho_o)\phi_{r\_ref} \tag{6.15}$$

From which the following can be deduced:

$$\cos\rho_o = \frac{q_1}{\sqrt{q_1^2+q_2^2}} \text{ and } \sin\rho_o = \frac{-q_2}{\sqrt{q_1^2+q_2^2}} \tag{6.16}$$

The expressions of sine and cosine functions can be introduced into the torque estimation expression (Eq. 6.12), resulting in:

$$\tilde{T} = p\frac{M^*}{L_r^*}\frac{\phi_{r\_ref}^2}{q_1^2+q_2^2}\left(\omega_r\frac{L_r}{M\,R_r}q_1 - \frac{1}{M}q_2\right) \tag{6.17}$$

Since the torque controller includes an integral action, under steady-state conditions, $\tilde{T} = T_{\text{ref}}$.

To determine the errors on flux and torque, a two-equation system is solved for each operating point determined by the values of Ω and T_{ref}. These equations are an expression of electromagnetic torque, depending on the reference flux and the real or estimated machine parameters (expression (6.17)), and the expression of magnetic saturation (2.137). Consequently, the slip pulsation is not calculated from the real torque as in Chaps. 4 and 5 but from the reference torque.

For the design of a torque control, the relative error on torque is particularly interesting. It is defined by:

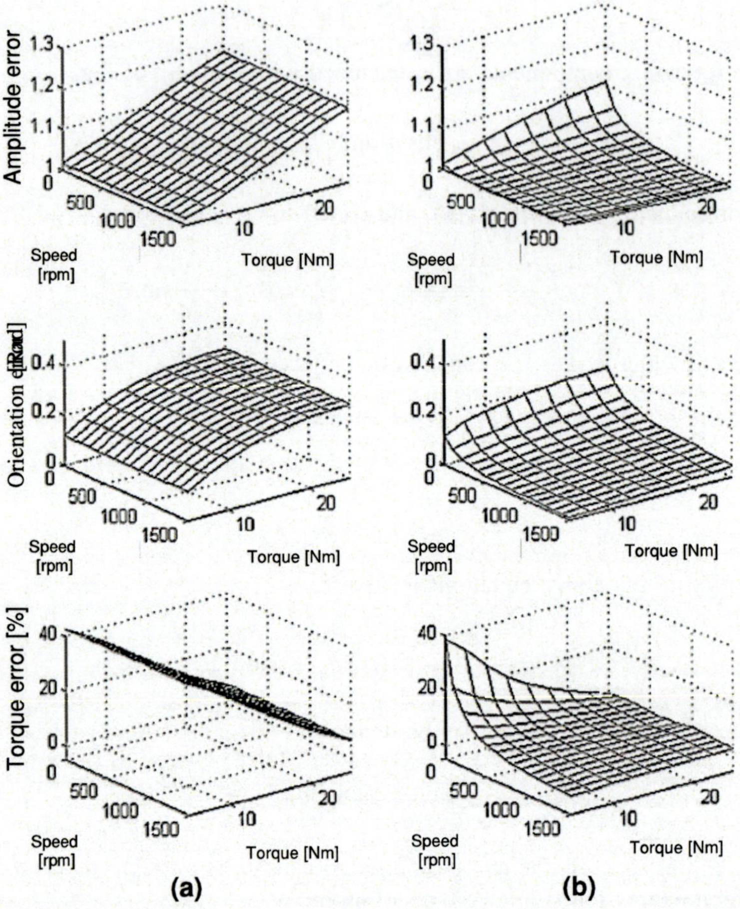

Fig. 6.19 Theoretical results of the reduced-order rotor flux observer with $R_r = 1.5R_r$*: **a** $k_1 = k_2 = 0$, **b** k_1 and k_2 are determined through fuzzy logic

$$\Delta T = \frac{\tilde{T} - T}{T} \tag{6.18}$$

Here, a 4,500 W induction machine has been used (see specifications at the end of this chapter):

Figure 6.19 represents the errors on the flux and on the torque (Eq. 6.18) when rotor resistance presents a 50% error. In Fig. 6.19a, the observer gains are null; in Fig. 6.19b, they are determined through the fuzzy logic supervisor. It was developed in Sect. 5.4 as a compromise between the different parameter sensitivities. A comparison between Fig. 6.19a and b shows that fuzzy logic helps to

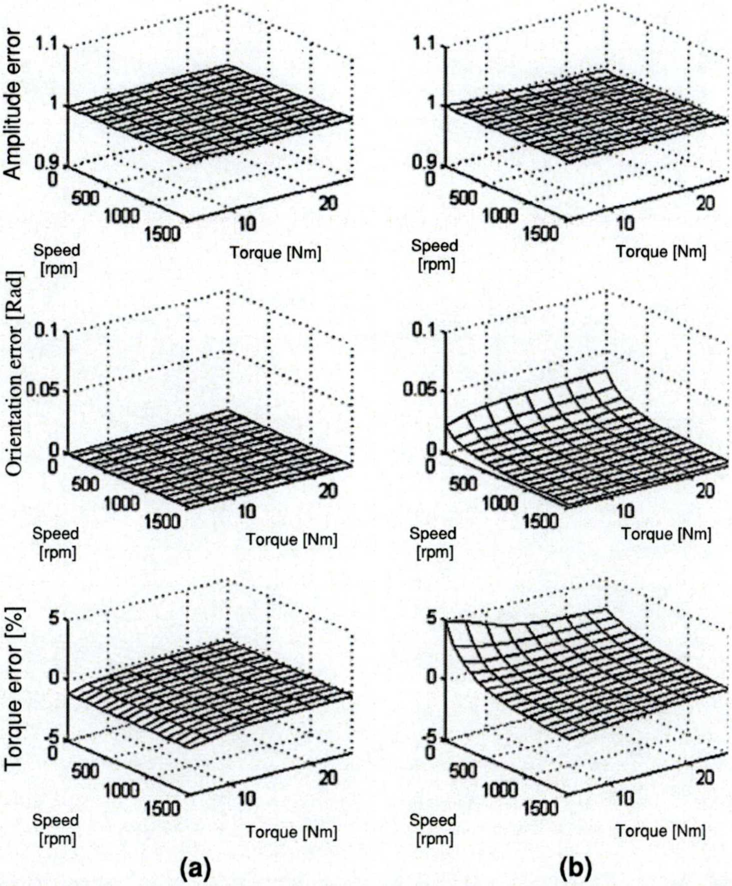

Fig. 6.20 Theoretical results of the reduced-order rotor flux observer with $R_s = 1.2R_s$*: **a**.$k_1 = k_2 = 0$, **b** k_1 and k_2 are determined through fuzzy logic

minimize errors on flux and torque. For a null speed and a torque of 25 Nm, the error on torque equals 6.7% for null gains and is down to 3% for gains determined through fuzzy logic.

Figure 6.20 presents the theoretical results with a 20% error on stator resistance. When the gains are null, the control is not sensitive to stator resistance. When the gains are not null, the system is sensitive to stator resistance, though the errors on flux and torque remain low.

Figure 6.21 presents the results obtained with a 20% error on mutual inductance. Once again, this figure illustrates the decrease in the number of errors on flux orientation and torque obtained with fuzzy logic.

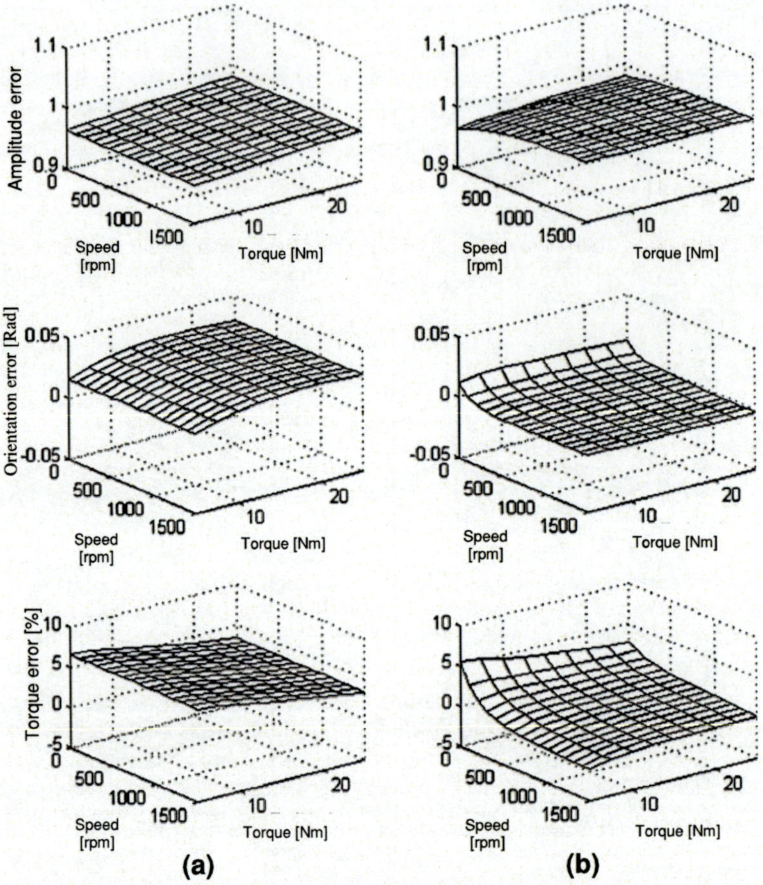

Fig. 6.21 Theoretical results of the reduced-order rotor flux observer with $M = 0.8M^*$: **a**. $k_1 = k_2 = 0$, **b**. k_1 and k_2 are determined through fuzzy logic

6.2.3 Results in Dynamic State

Figures 6.22 and 6.23 represent speed responses to a torque step, errors on flux, real torque (solid lines) and reference torque (doted lines) with a 50% error on rotor resistance. Figure 6.22 shows the results for $k_1 = k_2 = 0$. In Fig. 6.23, gains are determined through fuzzy logic. For the present studies, speed is kept null.

A comparison between Figs. 6.22 and 6.23 confirms the theoretical results: using fuzzy logic leads to optimum results and lower error rate on flux and torque.

While the error on torque reaches 23% in Fig. 6.22, it is only 10% in Fig. 6.23. However, it must be noted that the torque error remains high even when gains are determined with fuzzy logic. To further reduce the error, a parameter of the machine must be estimated (for example R_s) and a fuzzy supervision insensitive to R_r designed for the desired operating point.

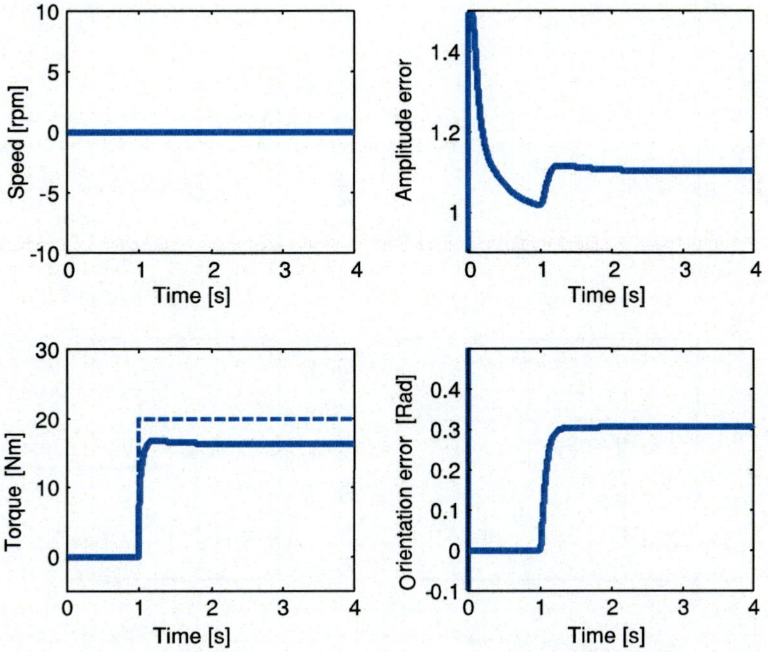

Fig. 6.22 Performance of the reduced-order rotor-flux observer $k_1 = k_2 = 0$ and $R_r = 1.5R_r$*

Figures 6.24 and 6.25 show the simulations realized with a 20% error on mutual inductance.

For the operating point studied here (null speed, maximum torque), the use of fuzzy logic reduce the error on flux orientation but increase the error on flux amplitude.

6.2.4 Parameters of the Tested Machine

Nominal power:	4.5 kW
Number of poles:	4
Stator resistance:	294 mΩ
Rotor resistance:	770 mΩ
Mutual inductance:	61 mH
Stator inductance:	65 mH
Rotor inductance:	65 mH
β:	0.7
s:	7.3

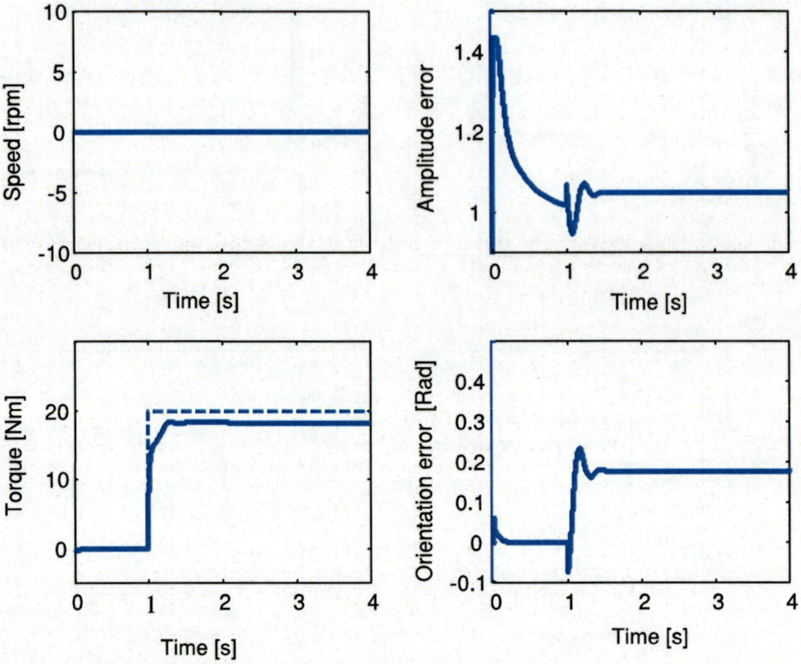

Fig. 6.23 Simulation of the reduced-order rotor-flux observer with k_1 and k_2 determined through fuzzy logic and $R_r = 1.5R_r^*$

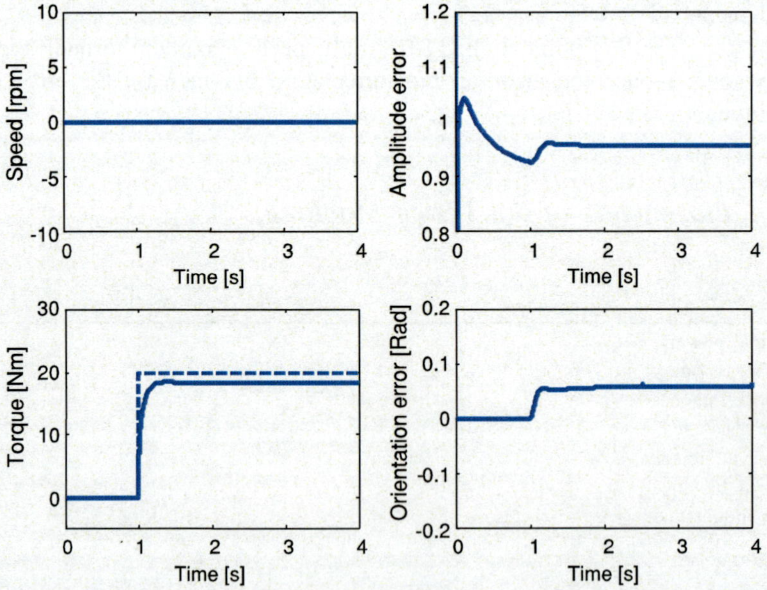

Fig. 6.24 Simulation of the reduced-order rotor-flux observer with $k_1 = k_2 = 0$ and $M = 0.8\ M^*$

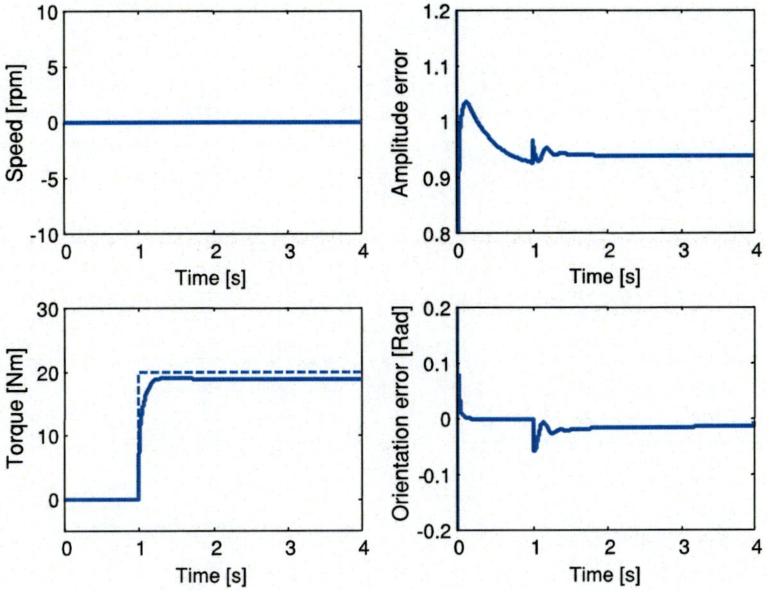

Fig. 6.25 Simulation of the reduced-order rotor-flux observer with k_1 and k_2 determined through fuzzy logic and $M = 0.8\ M^*$

References

Leclercq, L., Robyns, B., & Grave, J. M. (2003). Control based on fuzzy logic of a flywheel energy storage system associated with wind and diesel generators. *Mathematics and Computers in Simulation, 63*, 271–280.

Leclercq, L. (2004). *Apport du stockage inertiel associé à des éoliennes dans un réseau électrique en vue d'assurer des services système*. Ph. D. Thesus report from Université des Sciences et Technologies de Lille.

Cimuca, G. (2005). *Système inertiel de stockage d'énergie associé à des générateurs éoliens*. Ph.D. Thesus report from Ecole Nationale d'Arts et Métiers and Université Technique de Cluj-Napoca.

Cimuca, G., Saudemont, C., Robyns, B., & Radulescu, M. (2006). Control and performance evaluation of a flywheel energy storage system associated to a variable speed wind generator. *IEEE Transactions on Industrial Electronics, 53*(4), 1074–1085.

Berthereau, F. (2001). *Commande vectorielle multialgorithmique appliquée à la machine asynchrone avec optimisation par supervision floue*. Ph.D. Thesus report from Université des Sciences et Technologies de Lille.

Summary and Conclusion

In Chaps. 4–6, the sensitivity to uncertainties on electrical parameters of a number of control strategies have been analyzed. The different strategies that were studied are representative samples of commonly used vector control strategies.

As has been shown, all these strategies yield similar results in terms of parameter and dynamic performance. However, to obtain good results:

- Gains should be selected adequately,
- some of the estimated parameters must be judiciously used,
- several models must be combined,
- fuzzy logic should be used.

Yet these control strategies differ widely. Controls based on flux-observer data are indeed more complex to implement than when the flux component of axis q is supposed null. Moreover, the dimensioning and implementation of full-order observers is more difficult than that of a reduced-order observer.

Under these conditions, choosing a control strategy can be a real issue. Since the overall performance of all those control strategies are similar, does it make any sense to use the most complex one like those using flux observers?

Indirect control strategies are obviously the simplest. As has been shown, an integral action in current controllers can increase parameter sensitivity and PFF controllers are satisfactory, provided their gains are chosen correctly. However, with only one control parameter, sensitivity and dynamic performance are still closely linked. The performance of this control could be further improved with the development of a variable-structure current controller (Proportional Integral FeedForward or PIFF control). Indirect controls can be implemented with a relatively large sampling period (with the usual limits imposed by stability issues) and remain simple at the same time. Thus, their implementation only requires inexpensive digital CPUs with limited computation power.

The direct control assuming $\tilde{\phi}_{\text{rq}} = 0$, which was introduced in this book, is truly original. Combining two amplitude estimators for flux $\phi_r$with two flux orientation strategies (implicit and explicit feedback), made possible by the two coefficients

B. Robyns et al., *Vector Control of Induction Machines*, Power Systems, DOI: 10.1007/978-0-85729-901-7, © Springer-Verlag London 2012

K_ω and K_f, gives reliable control over parameter sensitivity and dynamics on the whole operating range of the machine. Coefficients K_ω and K_f are determined through the theoretical sensitivity study carried out with the help of fuzzy logic. Fuzzy logic is particularly helpful for this operation, as it requires managing the uncertainties brought about by the electrical parameter variations of the machine. The control strategy thus obtained remains simple and relatively independent from the sampling period, except for stability limits. Moderate processing power is therefore sufficient.

However, the flux estimator used in the strategy based on the assumption that $\tilde{\phi}_{rq} = 0$ cannot be put on an equal footing with a flux sensor. This estimator is indeed strongly dependant on the control strategy, especially the flux-orientation strategy (since it is assumed that $\tilde{\phi}_{rq} = 0$).

For a flux estimator to be considered equivalent to a flux sensor, it should logically estimate both flux components ϕ_{rd} and ϕ_{rq} without a priori. Flux observers could meet these needs. The question is whether an estimate of both flux components is really necessary. At this point of our study, the answer is yes for applications where the flux has an impact on other estimators:

- Electromagnetic torque estimator (railway traction systems for example);
- mechanical speed estimator (if the use of speed sensors is to be avoided).

If a flux observer is to have the performance of a real flux sensor, it must have good dynamics and low dependency on uncertainties for the parameters that intervene in the model of the machine used to develop it.

The methodology proposed in this book makes it possible to determine gains for flux observers ensuring a balance between dynamics, parameter sensitivity and observer simplicity. This method, which is based on the theoretical parameter sensitivity study, provides an observer which can be deemed optimum, as it is optimized for specific criteria such as dynamics, sensitivity, energetics and simplicity. This method has the advantage of potentially taking into account any number of criteria for the sensitivity study.

As a consequence, the optimized reduced-order flux observer developed in a stator-linked frame of reference with the proposed method can be considered similar to a flux sensor of which the input parameters would be mechanical speed, currents and stator voltages.

The computing time for vector controls based on a flux observer is higher than that of controls where $\tilde{\phi}_{rq} = 0$ is assumed a priori. More problematic is the fact that observers have a high sensitivity to the sampling period, which must consequently be reduced as much as possible. Digital processors for the implementation of flux observer controls must therefore have a rather high computing power, which implies higher costs of the control system.

Obviously, computing time increases along with the order of the observer. Although the optimum full-order observer developed in the stator reference frame can be considered equivalent to a flux sensor, it results in longer computing times.

It was however demonstrated that similar performance can be obtained with full and reduced-order observers (Robyns et al. 2000; Robyns 2000; Berthereau 2001; Berthereau et al. 2002). Then why use a full-order observer? The full-order observer introduced in this book delivers estimates for both components of the rotor flux and both components of the stator current. Although stator current is also being measured, estimated currents can prove useful for high power applications when the commutation frequency of the inverter semiconductors are limited to low values (maximum 1 kHz), thus increasing the length of the sampling period. As a matter of fact, a long sampling period results in delayed computations which in turn tend to destabilize current-control loops. Observer-estimated currents can partially compensate for these delays, as such currents foresee machine currents.

In conclusion, the choice of a control strategy depends on the application for which it is developed and on the resources available, which can be limited for budget or technology reasons.

The various techniques presented in this book for the analysis and development of controls help fine-tuning the choice of a vector control strategy for a given application.

References

Robyns, B., Berthereau, F., Cossart, G., Chevalier, L., Labrique, F., & Buyse, H. (2000). A methodology to determine gains of induction motor flux observers based on a theoretical parameter sensitivity analysis. *IEEE Transactions on Power Electronics, 15*(6), 983–995.

Robyns, B. (2000). Synthèse de commande robustes pour machine asynchrone basée sur une théorie caractérisant la sensibilité paramétrique. Research report from Université des Sciences et Technologies de Lille

Berthereau, F. (2001). Commande vectorielle multialgorithmique appliquée à la machine asynchrone avec optimisation par supervision floue (*PhD Thesus report from Université des Sciences et Technologies de Lille*)

Berthereau, F., Robyns, B., & Hautier, J. P. (2002). Fuzzy logic based gain determination of induction motor flux observers. *EPE Journal, 12*(2), 19–25.

Index

B. Robyns et al., *Vector Control of Induction Machines*, Power Systems,
DOI: 10.1007/978-0-85729-901-7, © Springer-Verlag London 2012